The Age of Manufactures
Industry, Innovation and Work in Britain
1700–1820

MAXINE BERG

Basil Blackwell
in association with Fontana

© Maxine Berg 1985

First published in this edition 1985 by
Basil Blackwell Ltd
108 Cowley Road, Oxford OX4 1JF, UK

First published 1985 by Fontana Paperbacks in
The Fontana History of England edited by G. R. Elton

British Library Cataloguing in Publication data

Berg, Maxine
 The age of manufactures: industry, innovation and work in Britain
 1700–1820.
 1. Great Britain—Industries—History
 I. Title
 338.4'767'0941   HC254.5

 ISBN 0-631-14639-3

Printed in Great Britain by T. J. Press Ltd, Padstow, Cornwall

For the economic history undergraduates at Warwick University, 1978 to 1984; and for Michael Lebowitz who taught me as an undergraduate, 1968 to 1970

Contents

Part Two   Paths to the Industrial Revolution

Preface

This book is about some aspects of 'the other Industrial Revolution', an 'Industrial Revolution' which included domestic industry and artisan workshops much more than it did the factory system; an Industrial Revolution which relied on tools, small machines and skilled labour much more than it did on steam engines and automatic processes; an Industrial Revolution which was created by women and children at least as much as it was by male artisans and factory workers. I have tried to make a case for a more long-run, varied and complicated picture of the British path to Industrial Revolution. And I have tried to do so, not by ignoring the centre points of traditional histories – cotton and iron, steam power and factories – but by placing these firmly in context with the experiences of the wide range of textile and metal industries, forms of work organization and technology.

Acknowledgements

I am grateful to friends and colleagues with whom I have discussed the ideas and themes of the book. My interest in the different methods of industrial and work organization originated in the work I did with Pat Hudson and Michael Sonenscher for our collaborative volume *Manufacture in Town and Country before the Factory* (Cambridge, 1983), and Chapter 3, 'Primitive Accumulation or Proto-industrialization?', is in part based on our joint introduction to that volume. Sara Mendelson, Ludmilla Jordanova, Pat Hudson, Michael Sonenscher, William Lazonick and Alec Ford read some of the chapters in draft, and I am grateful for their comments. Jeanette Neeson read the whole book in draft and provided me with very valuable comments and an important sense of perspective. Ruth Pearson gave me insight into the extent to which the 'other Industrial Revolution' is also a part of our own times, and just how much this revolution was and remains based on women's work. Hans Medick inspired my interest in seeking the connections between community and work organization. Peter Mathias, who originally suggested the book, was ever helpful and enthusiastic. Finally, Geoffrey Elton's suggestions were of great benefit in the revision of the book. I owe a great deal to his careful editing of the text. I am also grateful for the criticisms of an anonymous referee. All mistakes and omissions remain, however, my own.

Warwick University granted me two separate sabbatical terms when I wrote the bulk of the book, and the secretarial staff of the Economics Department typed the main draft. The Cambridge University Press granted me permission to write a revised version of 'Political Economy and the Principles of Manufacture 1700–1800' for Chapter 2. This was first printed in Berg, Hudson and Sonenscher, *Manufacture in Town and Country before the Factory*. The Ulster Folk and Transport Museum gave me permission to reprint the picture 'Flax Spinning in County Down', and Stanley Chapman and Christopher Aspin gave me permission to use the picture 'The Spinning Jenny' from their book *James Hargreaves and the Spinning Jenny* (Preston, 1964).

One of my tutors when I was an undergraduate, Michael Lebowitz, made history matter to me, and I am happy that some of my own students have caught this commitment. Another historian, John Robertson, has been my partner in work as much as in life, and our small daughters were both great gifts to us during the time in which I wrote the book.

M. B.

Introduction

The words 'Industrial Revolution' convey images of new technology and industry. Yet if we look at the many textbooks which have proliferated on the subject, we now find very little written directly about either technology or industry. Economic historians go to considerable lengths to define their 'industrial revolutions':[1] they have increasingly taken their definitions away from technology and industry, to concentrate instead on the phenomenon of economic growth. They have focused on the 'macroeconomics' of the Industrial Revolution, choosing to write about aggregate economic categories – patterns of economic growth, capital formation, demand, income distribution and economic fluctuations. They have rarely disaggregated their economy beyond the sectors – agriculture, industry, trade and transport. Their concerns have been those of economists who wrote in the 1960s of growth, development and capital investment. More recently, economic historians have turned their expertise towards presenting a progressively more quantitative Industrial Revolution; but as they have tended to shun areas without good statistical sources, assuming these to be the territory of the social historian,[2] they have not ranged beyond macroeconomic issues.

The 1980s brought a different economic climate, with a new concern among economists and their critics to question the results of the 'sacred cows' of the postwar boom – heavy capital investment, large-scale industry, new technology, structural change and rapid economic growth. Concerns have now changed to the structure of world recession, the causes and features of unemployment, and to the social and economic impact of new technology and new patterns of work organization. Our existing histories of the Industrial Revolution may present to some a story of past glories, but to many others they do not raise the interesting or relevant questions. Was rapid economic growth the experience of all regions of the country? Was there a great social divide between the employed and the unemployed, and just what did employment mean? How were new technologies introduced, and how did people react to them? How was industry organized and day-to-day work structured? These are questions which some social historians have tackled recently,[3] but the economic historians' Industrial Revolution has, on the whole, been untouched by such issues.

This must seem very odd to the layperson, for he or she would obviously ask where to find the bread and butter of the Industrial Revolution – the new technologies, the new industries, and the factory and domestic system. Of course some discussion of these lurks in most recent histories, but the only reasonably recent attempts to set down such a history have been Sidney Pollard's *Genesis of Modern Management* and David Landes's *Unbound Prometheus*. The scope, literary mastery and interpretive power of Landes's work are unsurpassed, and indeed, Landes's achievement has perhaps deterred other scholars from taking some of his questions further and from challenging his analyses. But it is also the case that Landes's Industrial Revolution suited its time. His Industrial Revolution was apocalyptic; his vision of its processes cataclysmic. It fitted into contemporary studies of economic growth; for his concerns with the achievements of the factory and large-scale power technologies confirmed contemporary approval of heavy capital investment. Landes's work also informed the concerns of social historians at the time, making the grievances of factory workers and the social conflict of the 1830s and 1840s the great area of historical debate.

In the 1960s Landes could write, 'The labouring poor, especially those by-passed or squeezed by machine industry, said little but were undoubtedly of another mind.' Today this is no longer adequate. Today we want to know about the social implications of technological change, not only in our own time, but in the past. The failings of our own heavily capitalized, large-scale factories have been measured against the revival of other alternative smaller-scale units. And the inflexibilities and industrial conflict of hierarchically organized systems of management have entailed new attempts at cooperative production and decision-making.

These questions about our own time have created a need for a new microeconomics of the Industrial Revolution: the forms of industrial organization not just in the factory system, but in the putting-out system, artisanship, subcontracting, and mining organization; the characteristics of the labour force, forms of recruitment and industrial training; and the types of technology – traditional as well as new, manual as well as power, small-scale and intermediate change as well as large-scale; and the varied experiences of industries and regions – experiences of industrial decline as well as of growth. These issues challenge us to a much more open-ended study of the Industrial Revolution: we need to study the controversies and conflicts underpinning change, not just their results in the indices of economic growth; and we need to study the failures as well as the success stories, for these were equally a part of the experience of industrialization.

Our own current Western European experience of industrial growth and recession, coupled with the recent rise of manufacturing in many Third World countries, has also raised challenging questions about the meaning of industrialization and the form which it takes. The old aspirations for large-scale, capital-intensive factories and mechanization have given way before new, small-scale information technologies, before a new decentralization and 'new international division of labour',[4] and the possibilities of 'alternatives to mass production'.[5] We now look on industrialization as a cyclical process rather than as a one-way progression, as long-term rather than as short-term and dramatic, and as multidimensional rather than as a single model.

Anthropologists and development economists have been particularly struck in recent years, not by the similarities between the new Third World manufacture (particularly manufacture which takes place in the so-called 'informal sector') and the full-scale European industrial revolution, but by its similarities with pre-industrial conditions and the transition years before industrial revolution.[6] It is the historical question mark over the eventual outcome of 'proto-industrialization', that is, the rise of cottage manufacture and the putting-out system, which underlies uncertainties over the future of small-scale industry and other forms of manufacture in the Third World today, even though the world context for such manufacture is so very different.

This book takes up these questions of types of technological change and forms of industrial organization for the study of the Industrial Revolution itself. The long and varied experience of organizational and technological change was an experience which went back to the early eighteenth century as much as forward to the early nineteenth. But recent textbooks concentrate only on the period after 1780, leaving the rest of the eighteenth century as an appendage to pre-industrial history. Here we will look at the whole of that century, not just at its last dramatic decades.

My interests in the earlier phases of the Industrial Revolution, in the long course of industrialization and in questions of technology, work organization and regional social and institutional change are not, however, new interests unique to our own time. T. S. Ashton, Paul Mantoux and Charles Wilson, writing about the whole of the eighteenth century, did dwell upon the technological and industrial side of the Industrial Revolution,[7] but they found their framework within a much older tradition, one which went back in the first instance to the 1920s and 1930s, and to some extent even further back to the historical economists and economic historians of the first years of the twentieth century.

Industrial history was then a terrain of controversy for socialists and their critics, for the socialists were profoundly interested in noncapitalist forms of organization, and in the origins of capitalism and wage labour. A. P. Usher framed his monumental *An Introduction to the Industrial History of England* in terms of a response to the rising tide of socialism in the years just after the First World War. He opened his book with a critique of socialist economic history, in particular that of the German socialist Rodbertus. The interest in industrial organization was also an aspect of both the rise of an economic interpretation of history, and of an historical economics. Attempts were made to define and analyse the forms of industrial organization – guild structures, domestic or cottage manufacture, and factory production. These attempts to find historical systems of economic activity subsequently fell from fashion, but they did at least wield an important influence over the intensive and scholarly work of economic historians from the First World War to the 1930s. These delved in depth into the industrial history, work practices and technologies of pre-industrial manufacture and the early Industrial Revolution.[8]

There is thus a long tradition of special interest in the structures, work practices and labour forces of household production units and the putting-out system. The Industrial Revolution and the factory system were placed firmly in the historical perspective of the long genesis of industry. Historical investigation of the different forms of manufacturing, of working conditions, of the special features of female and child labour was integral to controversy over optimistic and pessimistic interpretations of industrialization. We hear frequently of J. H. Clapham's optimistic response to the Hammonds.[9] We hear less of the work of an important group of women historians at the time – Alice Clark, Ivy Pinchbeck and Dorothy George – to dispel the myths of a golden age around seventeenth- and eighteenth-century industry, fabricated upon the domestic system and the women's and children's labour which dominated this industry.[10] The ubiquity and comparative success of domestic and workshop manufacture in the eighteenth century, and its continuation alongside the factory system well into the nineteenth, were based on the intensive exploitation of labour, particularly that of women and children, an exploitation at least equal to that suffered under the factory system.[11]

In our own time when the attractions held out by small-scale and decentralized industry, and by small-scale and labour-intensive technologies, seem a welcome relief from the factory and machinery, we need to look anew, from an equally critical historical perspective, at the ways in which patterns of work and technology were used in the past. Small was

sometimes beautiful, but more often it was dependent, op-
pressive and exploitative. For the factory and the domestic
system, power technologies and manual skills, male artisans as
well as women's and family labour were all used either as
alternatives or together, depending on time and industry, within
a wider system of prices and profits.

This book challenges the present singular attachment among
economic historians to the years after 1780, to the factory and to
the cotton industry. It asks that we reconsider the kinds of
changes taking place in the earlier eighteenth century and the
context in these years for the rise of household and workshop
industries. It asks for a closer analysis of the economic dynamic,
the techniques and the labour forces, of these cottage and
workshop industries and of the factories which grew up in the
midst of some but not all of them. It asks finally that we look on
the Industrial Revolution as a more complex, many-sided and
long-term phenomenon than economic historians have recently
assumed.

This book raises a number of areas of debate and analysis, but
it does not by any means provide the kind of broadly based
industrial history we now need. It is necessarily selective,
dealing in depth with only some of the textile industries and
some of the metal industries. As a history of manufacture, it
deals with only the two major categories of manufacture at the
time, and leaves untouched a whole series of smaller but very
important manufacturing industries. As a general survey, it
raises more questions than it answers; one of these questions,
which is dealt with at only a very superficial level, concerns the
impact on and response of the workforce in the eighteenth
century to the introduction of new techniques and work
practices. Our knowledge of this is still far too limited. There
are many other gaps both in the general discussion and in the
detailed history of the book. But I hope these will provide the
incentive for new research and new interpretation on the
eighteenth-century economy.

PART ONE

Manufacture and the Economy

CHAPTER 1

Industries

The British economy in the eighteenth century contained all the contrasts of youth's inclinations, at once the backward and the forward, its gaucheness as well as its quickness of perception and response. The 'age of improvement', to which we now look back, was cast within an 'age of manufactures' where the skills and traditions of artisans mingled with new products, markets, labour forces and, above all, mechanical contrivances. The new technologies and the spirit of innovation which mark out this century were alarming in their powers of transformation, and yet apparently easily absorbed into existing social structures. It was a century to which Clapham could attach the label of an 'Industrial Revolution in slow motion', but to which Landes could also pin his title, 'Unbound Prometheus'.

If we look at some of the economic indicators of development in the eighteenth and first years of the nineteenth century, we find that it is in practice very difficult to identify precisely the economic origins and effects of most of the large- and small-scale innovations which transformed this world. On the one hand, we have Landes's description of the bold sweep of technology's advance. It came like a republican army, confident in its principles and its moral precepts, and thus overcoming all economic, social and historical barriers. It grew, too, as did a new radical and anti-establishment movement, by finding new geographical frontiers to practise new technologies and new industries. Landes's Industrial Revolution, whose pre-ordained triumph was so obvious and so right, was based on the achievements of new machinery, new power sources and new raw materials.[1] His picture was centred on the cotton industry, reinforced by the development of steam power and the expansion of the metal industries. As Landes himself put it, 'The cotton mill was the symbol of Britain's industrial greatness; the cotton hand her greatest social problem – the rise of an industrial proletariat.'[2] It is true that Landes did give some credence to that other classic, Clapham's 'Industrial Revolution in slow motion', but where Clapham had emphasized the continuities of the period Landes's vision was unashamedly apocalyptic. His study of the new technology and the factory system was not that of some 'gentle erosion of the traditional order', for 'the people of the day were not deceived by the pristine air of much of Britain's landscape. They knew they had passed through a revolution.'[3]

Clapham's view of the slow rate at which the factory system and new techniques spread across British industry was set down in the 1930s but it now bears revival and a new interpretation. Then Clapham pointed out the persistence of traditional forms of organization and labour-intensive techniques well after the age of the factory and steam-powered machinery. Now we can attempt to explain this persistence by the different microeconomies of the various sectors and industries, and by the regional and cyclical pattern of industrialization.

Industrialization showed a very uneven pattern over both regions and sectors of the economy. Eighteenth-century British industries grew as much within the old framework of home and artisan manufacture as within the more noticeable new factories and mechanized works. For much of the eighteenth century they also grew rather slowly. The early phases of industrialization appear to have had very little impact on aggregate growth rates. In fact, the annual average rates of growth in agriculture and industry differed little between Britain and France for much of the eighteenth century. The productive capacity of the British economy was, furthermore, exceeded by that of France right up until the French Revolution.

It was only by the early years of the nineteenth century that Britain's economic superiority was clear. And between 1810 and 1900 her total commodity output grew at an average rate of 2.6 per cent per year; France's grew by only 1.5 per cent. Over this period Britain's population rose threefold, while France's

Table 1. *Annual average rates of growth, 1701–10 to 1781–90 (percentages)*

	Agricultural production	Industrial production	Total commodity output
Britain	0.4	1.1	0.7
France	0.6	1.9	1.0

increased by only one third. Increases in output per head over the same period were thus much closer for the two countries: Britain's per capita rate of growth in commodity output was 1.3 per cent while France's was 1.2 per cent. This comparison of economic growth in Britain and France relies on Deane and Cole's estimates for Britain, estimates which have recently been revised downward. Britain's position was probably even closer to France's for the late eighteenth century, and even for the early nineteenth the divergence in growth, while apparent, was not so startling.[4]

Comparing the two countries after considering differences in population does seem to play down the extent of Britain's economic transformation. But that transformation was not an illusion, especially if we rely on traditional estimates. For rates of growth in total production in the late eighteenth and nineteenth centuries ran far ahead of those for the rest of the eighteenth century. Total production grew by 1.7 times between 1700 and 1780, and production per capita by 1.2 times; but in the years 1780–1881 total production grew by 12 times and production per capita by 3.5 times.[5]

If we rely on aggregate data – figures for total production, total population, total agricultural output – most of the eighteenth century appears as a period of little change. More significant change obviously took place at the end of the eighteenth century and in the nineteenth. But these figures should not allow us to magnify the late eighteenth-century transformation out of all proportion.

Recent estimates have, in fact, given an impression of slower but steadier and more widespread growth than we have usually assumed. First, it is now argued that per capita growth in the years traditionally assigned to the Industrial Revolution was really very modest. Annual growth in per capita income has been estimated at only 0.33 per cent between 1770 and 1815. The picture of economic growth starts to look better only when a substantial part of the nineteenth century is included in estimates. Second, it is clear that the big three of the Industrial Revolution – cotton, iron and engineering – accounted for only a small part of industry. They produced less than one quarter of British manufactures even in the 1840s, and the older textile industries, agricultural processing (milling, baking, brewing and distilling) and leather processing generated more income than the technological leaders. Total productivity grew rapidly among the leaders, but it was by no means static in other industries. From 1780 to 1860 total productivity in the large modernized sectors of textiles, canals, railways and shipping rose by 1.8 per cent per year; while in other and nonmodernized sectors such as commerce, clothing, food processing, mechanical implements, domestic service, building, the professions, chemicals, potteries, glass, gasworks, tanning and furniture making it rose at 0.65 per cent. The differences at first sight seem stark. But productivity increases of 0.65 per cent per year are no mean achievement. They indicate 'that ordinary inventiveness was widespread in the British economy 1780 to 1860 . . . the Industrial Revolution was not the Age of Cotton or of Railways, or even of steam entirely; it was an age of improvement'.[6]

Revisionists have furthermore argued that the industrial sector was in the eighteenth century 'nearly twice as large as previous estimates indicated' and that 'its subsequent transformation was less dramatic'.[7] But unexciting as this transformation may then appear, by disaggregating some of our indicators of growth we immediately magnify the experience of each industry. Casting the net wider than the industrial big three does not so much detract from their success as enhance the wider context of industrial improvement within which they prospered. Looking back to earlier decades of the eighteenth century with more interest in the growth they did achieve, we can place industry in

a more historical context of the long eighteenth century than
that of the short spurt at the end of that century. Table 2
indicates a trend of increase in real output over the whole of the
eighteenth century, but especially from 1740.

Table 2. *Index numbers of eighteenth-century real output (1700 = 100)*

	1700	1720	1740	1760	1780	1800
Industry and commerce	100	105	131	179	197	387
Agriculture	100	105	104	115	126	143
Rent and services	100	103	102	113	129	157
Government and defence	100	91	148	310	400	607
Total real output	100	108	115	147	167	251
Average real output	100	105	113	130	129	160

The origins of higher growth rates in most eighteenth-
century industries go back at least as far as the 1740s. Deane
and Cole's estimates of increases in output emphasize the existence of a clear trend upwards from at least this time. These
estimates have, in turn, been subject to further scrutiny which
has pushed growth back to earlier decades in the eighteenth
century, called into question the existence of an upturn after
1740 and smoothed the upturn after 1780. The following table
shows this smoother, more limited growth path entailed by
Crafts's estimates, and there is other evidence of higher
agricultural output at the beginning of the eighteenth century
than Deane and Cole allowed for.

Such wide categories as these, however, tell us nothing about
the experience of individual industries. How did the linen industry fare as against cotton, or copper and brass as against iron?
We must look in more depth at the industrial groups which form
the staple fare of the Industrial Revolution – textiles and metals
and mining. But we must also look at indicators of the course of

Table 3. *Estimates of growth in industry and commerce (per cent per annum).*

	Deane and Cole	Crafts
1700–60	0.98	0.70
1760–80	0.49	1.05
1780–1801	3.43	1.81
1801–31	3.97	2.71

change in traditional pre-industrial trades and manufactures –
building, brewing, candlemaking, papermaking, starch- and
soapmaking, the leather trades and glass manufacture.

Eighteenth- and early nineteenth-century growth rates across
a range of industries have been summarized recently by Crafts

Table 4. *Sectoral growth rates of real output (per cent per annum)*

	Cotton	Wool	Linen	Silk	Building	Iron	Copper
1700–60	1.37	0.97	1.25	0.67	0.74	0.60	2.62
1760–70	4.59	1.30	2.68	3.40	0.34	1.65	5.61
1770–80	6.20		3.42	−0.03	4.24	4.47	2.40
1780–90	12.76	0.54	−0.34	1.13	3.22	3.79	4.14
1790–1801	6.73		0.00	−0.67	2.01	6.48	−0.85
1801–11	4.49	1.64	1.07	1.65	2.05	7.45	−0.88
1811–21	5.59		3.40	6.04	3.61	−0.28	3.22
1821–31	6.82	2.03	3.03	6.08	3.14	6.47	3.43

	Beer	Leather	Soap	Candles	Coal	Paper
1700–60	0.21	0.25	0.28	0.49	0.64	1.51
1760–70	−0.10	−0.10	0.62	0.71	2.19	2.09
1770–80	1.10	0.82	1.32	1.15	2.48	0.00
1780–90	0.82	0.95	1.34	0.43	2.36	5.62
1790–1801	1.54	0.63	2.19	2.19	3.21	1.02
1801–11	0.79	2.13	2.63	1.34	2.53	3.34
1811–21	−0.47	−0.94	2.42	1.80	2.76	1.73
1821–31	0.66	1.15	2.41	2.27	3.68	2.21

to show that though there was rapid growth and structural change towards cotton and iron at the end of the eighteenth century, there was also substantial growth across a range of traditional industries including wool, linen, silk, copper, coal and paper before 1770.[8]

Not only was the industrial sector much larger and its growth more widespread than we previously believed, but the occupational structure of eighteenth-century England was much more industrial. The traditional surveys used by historians to tell us about the distribution of employment in pre-industrial and early industrial England are those of Gregory King (1688), Joseph Massie (1759) and Patrick Colquhoun (1811). Recent

Table 5. *Occupations in eighteenth-century England*

COMMERCE AND INDUSTRY

King		Lindert	
Merchants & Traders by Sea (greater)	2,000	All Commerce	135,333
Merchants & Traders by Sea (lesser)	8,000	Manufacturing Trades	179,774
Shopkeepers & Tradesmen	40,000	Mining	15,082
Artizans & handycrafts	60,000	Building Trades	77,232
	110,000		407,421 (excluding labourers)

AGRICULTURE

King		Lindert	
Freeholders (greater)	40,000	All Agriculture (excluding labourers)	241,373
Freeholders (lesser)	140,000		
Farmers	150,000		
	330,000		

econometric estimates based on burial records give much greater emphasis than did King to industrial occupations, and these offer in addition some breakdown of occupations. Such estimates, to be sure, come with large margins of error. Even allowing for these, the new Lindert estimates do convey a much more industrial Britain than we have previously assumed. Lindert argues, furthermore, that though occupational structure was relatively stable before 1755, the two groups which expanded faster than average were agriculture and manufacturing. During the period of the industrial revolution manufacturing employment increased, dominated by the textile trades. Estimates show that employment in textiles more than tripled in the second half of the eighteenth century, and the number of weavers more than doubled. Other rapidly increasing occupational groups included building and mining.[9]

Let us now look at the pattern of growth within some individual industries.

Textiles

The textile group of industries was a complex combination of the old and the new, of handicraft and factory industry. Through the eighteenth century and even the early nineteenth both sides expanded. The fortunes of the traditional woollen industry, which did not experience the spectacular technical and organizational transformations of cotton, were still sufficient for it to dominate over the cotton industry until the 1820s. The growth of this last was indeed amazing. Deane and Cole show that raw cotton imports, which increased by a third in the early eighteenth century, doubled between 1750 and 1775 and then increased eightfold in the last twenty years of the century.[10] Landes contrasts the two periods of the cotton industry's expansion by pointing out that in 1760 Britain imported only 2½ million pounds of raw cotton to feed an industry dispersed through the countryside; but in 1787 she consumed up to 22 million pounds, the industry was second only to wool in numbers employed and value of production, and most of the fibre was being cleaned, carded and spun on machines. For this was the period of the great textile innovations – Kay's flying shuttle, Hargreaves's jenny, Arkwright's water frame and Crompton's mule, to be followed at the end of the century by the power loom and dressing frame.[11]

If we compare the course of change in the cotton industry with the woollen, linen and silk industries, we see that the output of these industries increased substantially in the eighteenth century, especially from mid-century. The woollen industry expanded steadily in the first forty years of the century at a rate of about 8 per cent per decade, but in the next thirty years grew by 13 to 14 per cent, and then dropped to about 6 per cent in the last quarter of the century.[12] The course of change in the cotton and woollen industries is shown in Table 6.

Behind this trend of change in the woollen industry lay much more remarkable changes in its geographical distribution. In the early eighteenth century the geography of the industry was much as it had been in the Middle Ages, that is to say, concentrated in East Anglia, the West Country and Yorkshire. But the rapid growth of the worsted industry in Yorkshire from the late seventeenth century was reflected in the eighteenth-century growth of Leeds, Bradford, Huddersfield, Wakefield and Halifax. The new importance of the West Riding of Yorkshire was complemented, however, by the decline in the course of the century of the Suffolk, Essex and West Country woollen industries.[13] And the woollen industry of Lancashire was encroached upon by the fustian industry (cotton and linen mix) even from the seventeenth century.

If the progress of the woollen industry was so substantial, particularly in the mid-eighteenth century, that of two other traditional textile industries seems equally impressive. The silk industry, in spite of its handicap as an expensive luxury commodity facing fierce foreign competition, was a technological leader from the first quarter of the eighteenth century in the use of water-driven machinery and the factory system. Imports of raw silk fluctuated through the first half of that century and thereafter increased steadily from 670,000 pounds in 1750–9 to

Table 6. *The cotton and woollen industries, 1695–1824*[14]

	Retained imports Raw cotton million lbs.	Wool consumed (including imports) million lbs.	Value Added Cotton £ million	Value Added Wool £ million	Gross Value Value Final Product Cotton £ million	Value Final Product Wool £ million[15]
1695–1704 (1695)*	1.14	40	—	3.0	—	5.0
1740–9 (1741)	2.06	57	—	3.6	—	5.1
1772–4 (1772)	4.2	85	0.6	7.0	0.9	10.2
1798–1800 (1779)	41.8	98	5.4	8.3	11.1	13.8
1805–7 (1805)	63.1	105	14.4	12.8	18.9	22.3
1819–21 (1820–4)	141.0	140	23.2	16.6	29.4	26.0

*Dates between brackets are for wool.

Table 7. *Change in the linen and silk industries*

IMPORTS OF FLAX, YARN AND SILK[16]

	Flax imports England (cwt.)	Yarn imports England (cwt.)	Flax imports Scotland (cwt.)	Raw and thrown silk imports Britain (000 lbs.)
1700 (1700–9)	62,701	17,921	—	499
1720 (1720–9)	37,310	27,458	—	675
1740 (1740–9)	69,572	27,071	686	552
1760 (1760–9)	73,059	62,537	3,325	906
1780 (1780–9)	146,734	91,914	4,777	1,132
1790 (1790–9)	145,056	79,855	7,045	1,181

LINEN STAMPED FOR SALE IN SCOTLAND, 1728–1822[17]

	Million yards	Thousand £
1728	2.2	103
1748	7.4	294
1768	11.8	600
1788	20.5	855
1808	19.4	1,015
1822	36.3	1,396

APPROXIMATE LINEN EQUIVALENT OF ENGLISH FLAX AND LINEN YARN IMPORTS[18]

	Thousand yards
1700	12,393
1720	12,427
1740	16,375
1760	26,803
1780	44,343
1790	40,735

Table 8. *The textile industries in the eighteenth century: comparison of raw materials consumed*[19]

	Retained imports Raw cotton million lbs.	Wool consumed million lbs.	Flax imports England and Scotland million lbs.	Linen yarn imports million lbs.	Raw and thrown silk imports million lbs.
1695–1704	1.14	40	6.3	1.8	0.5
1740–9	2.06	57	7.0	2.7	0.55
1760–9	3.53	—	7.6	6.2	0.9
1772	4.2	85	12.5	9.5	0.95
1780–9	15.51	—	15.1	9.1	1.1
1798–1800	41.8	98	15.2	7.9	1.2

1,181,000 pounds in 1790–9.[20] The linen industry showed remarkable expansion in the second half of the century, in spite of a slackening and even decline in the imports of flax and linen yarn. Linen had long been a widespread local industry for domestic consumption in England; in the course of the eighteenth century it became much more market-orientated. The output of English linen doubled in the second quarter of the eighteenth century and did so again in the third quarter. In the more concentrated commercial linen industry of Scotland, the value of linen produced rose from £103,000 in 1728 to £1,116,000 in 1799, and fluctuated thereafter until the 1820s. A comparison of the course of change in the linen and silk industries is shown in Table 7.

The course of economic growth in the eighteenth-century textile industries was indeed overshadowed by the spectacular growth and regional concentration of the cotton industry. But we should not forget the still predominant position of the woollen industry and the steady growth of the other textile industries.

Mining and metals

The fuel which was the key to the British path of technical transformation was coal. A rise in the output of coal can be dated from the late 1740s, and between 1760 and the 1780s output rose from 5 to 10 million tons. But even earlier in the eighteenth century output had risen steadily from 3 million tons in 1700 to 5 million in 1760.[21] This increase was a response to new demands for coal as a domestic fuel as well as for its use in lime kilns, and metalworking. It also reflects a growing substitution of coal for wood in the manufacture of salt, sugar and soap, in the making of glass, starch and candles, in brick- and tilemaking, in dyeing and brewing, and finally in the smelting and forging of iron.[22] The huge increases in the output of coal, particularly in the last half of the century, were achieved by greater use of labour power, for the industry was largely unmechanized until after 1914. The major change in the industry in the eighteenth century was the sinking of deeper shafts and the use of steam engines, mainly the Savery and the Newcomen engines, to drain them.

In metal smelting and refining the major technical changes of the period were those that saved on raw material costs, not on labour. The shift from charcoal to coke smelting of iron had massive repercussions. Deane and Cole found the growth of the iron and steel industries to be relatively slow before 1760. Between 1757 and 1788 the rate of increase was barely 40 per cent per decade. But between 1788 and 1806 decennial rates of increase rose to over 100 per cent. Imports of bar iron, the raw material of the new metalworking industries, trebled between 1700 and 1800, and in the short period 1755–64 to 1790–9 they rose from 33,000 tons to 49,000.[23] Britain's own production of iron also pushed forward, and exports of iron and steel swelled from 1,600 tons in 1700–9 to 14,300 tons in 1790–9.[24]

It took some time for the new coke-smelted iron to make any serious impression; old-style charcoal smelting remained popular for a long time. Darby first smelted iron with coke at Coalbrookdale in 1709. In 1717 John Fuller counted 60 charcoal furnaces and 144 forges, most near supplies of iron ore, scattered through the Weald, Yorkshire, Derbyshire, Staffordshire, Shropshire, Monmouthshire and South Wales. But even by 1774 there were only 31 coke furnaces in blast in Britain. It was only after this time that ironmasters in force made the change. By 1790 there were 81 coke furnaces as against only 25 charcoal furnaces. The output of pig iron accordingly increased nearly four times between 1788 and 1806.

Table 9. *Output of pig iron, 1717–1806 (tons)*[25]

	England and Wales	Scotland
1717	18,000	—
1788	61,000	7,000
1796	109,000	16,000
1806	235,000	23,000

1775 was the beginning of a new era when the wider application of steam power made for a stronger blast and raised the efficiency of smelting by coke. It also witnessed the geographical shift of the industry to the Midlands, Yorkshire, Derbyshire, South Wales and Scotland. The effect of this new technology may be gauged by the fact that whereas in 1750 Britain imported twice as much iron as she made, in 1814 she exported five times as much as she bought.[26]

As occurred in the relation of cotton to the other textile industries, so the remarkable progress of the iron industry was accompanied by the sustained expansion of other traditional nonferrous metal industries. The exports of these metals more than quadrupled over the century – from £232,000 up to £1,160,000 in fixed prices.[27] Cornish tin production rose from 1,323 tons in 1695–1704 to 2,658 in 1750–9 and increased still more to 3,245 tons in 1790–9. Cornish copper-ore production rose from 6,600 tons in 1725–34 to 46,700 tons in 1790–9.

The new technology in iron processing and the increase in output in the nonferrous metal industries brought about a shift to a coal fuel and metal-based technology. The diffusion of coke smelting had also put pressure on the old refining techniques. Puddling and rolling replaced them and made it possible to substitute for the expensive imported bar iron, traditionally used in metalworking, the new home-produced pig iron.

It was not just the production of basic metals which made great advances, but that of all sorts of metal goods. Well before the Industrial Revolution the West Midlands, South Yorkshire and northeast Durham were well-established centres of the metal trades. In 1677 Andrew Yarranton described how the iron manufactures of Stourbridge, Dudley, Wolverhampton, Sedgley, Walsall and Birmingham were diffused all over England. The Crawley ironworks in northeast Durham were established in 1682 in Sunderland, moving in 1690 close to Newcastle. And Defoe wrote in 1724 of 'the continued smoke of the forges' at Sheffield which were always at work, making all sorts of cutleryware.[28]

The new proliferation of metalworking was founded not on machines, but on labour and skill. The puddling and rolling of

iron were based on the skills and the sweat and muscle of newly
specialized workers – the furnaceman and the puddler. The emergence of this freshly skilled workforce arose in turn out of a long
tradition of artisan skills in mineral fuel technology. Metalworking
covered not only the hardware and cutlery trades, but machine- and
toolmaking. The metalworking trades, based by and large on
artisans working with rudimentary tools such as files and grinding
stones in small workshops, were improved by new handtools, for
turning, cutting, piercing and stamping metal goods. Improvements
in water power and the introduction of an effective rotative, then
rotary, steam engine, while they did not immediately affect most of
these small-scale trades, did demand new millwrighting and engineering skills, and especially skills in the precision metalworking of
large objects in iron, such as cylinders, crankshafts and piping. And
a new range of engineering tools for boring, planing, turning and
cutting accompanied these skills.

There is no way of estimating quantities of new tools and
machinery produced. Although exports were prohibited, much
machinery was smuggled abroad. Neither are there any indices
for the home trade. Much of this was made in machine shops
attached to factories or works, or in shops producing a range of
engineering and hardware goods. The only major centres of
engineering technology by the early nineteenth century were
Boulton and Watt's Soho Works, Maudslay's London Workshop, and the Woolwich Arsenal.

Other industries

Textiles and metals were the technological leaders, but output also
increased in a whole series of industries whose technologies were
based on traditional handicraft work or new skills – leathermaking,
shoemaking, glove- and hatmaking, even pinmaking, woodworking
and the building and construction industry.

Apart from the woollen manufacture, the most important
eighteenth-century industries were leather and building. The
contributions of these and other 'traditional' sectors to industrial
output is illustrated in Table 10.

Table 10. *Value added in industry (£ million current)*[29]

	1770	1801	1831
Cotton	0.6	9.2	25.3
Wool	7.0	10.1	15.9
Linen	1.9	2.6	5.0
Silk	1.0	2.0	5.8
Building	2.4	9.3	26.5
Iron	1.5	4.0	7.6
Copper	0.2	0.9	0.8
Beer	1.3	2.5	5.2
Leather	5.1	8.4	9.8
Soap	0.3	0.8	1.2
Candles	0.5	1.0	1.2
Coal	0.9	2.7	7.9
Paper	0.1	0.6	0.8
	22.8	54.1	113.0

The leather industry contributed over 23 per cent of total value added in industry, only a little less than the 30 per cent contributed by wool. Building contributed 10 per cent. While the value added by the leather industries expanded substantially by 1801, though less so by 1831, its share of total value added in industry fell to less than 10 per cent. The value added by building continued to expand both absolutely and proportionately over the whole period. Its share of total value added in industry rose to 17 per cent in 1801 and further to 23 per cent in 1831. Substantial absolute and proportionate increases in value added were also apparent in a range of other traditional industries. The contribution of coal increased steadily from 2 per cent to 6 per cent of industrial value added; likewise the values added by soap and beer increased absolutely and at least maintained their relative share. The growth rates of real output for all of these traditional industries were respectable, and in some cases very substantial. In the cases of building and coal, these rates rank only behind those of cotton and iron.

Such large increases in output almost certainly entailed some change in the organization of production, if not in actual technical processes. Certainly expansion involved the much larger input of labour, but marshalling the forces of greater numbers of workers must have involved organizational change which probably affected the division of labour. Some of these industries have been the subject of industrial and business histories,[30] but in the case of many of them we know little about their labour forces, work processes and organization during these early phases of industrialization. They have not formed a part of the traditional fare of the Industrial Revolution, and their absence is clearly mistaken.

Changes in organization

The increases in total productivity which form the basis of higher output figures across eighteenth-century industry were achieved by both technical and organizational change. New technologies – machines, tools and human skill – cannot be separated from the context in which they were used. Potential for higher output through such new techniques could be enhanced on the one hand, or limited on the other, by the organization of the production process. A more effective division of labour, the intensification and greater exploitation of the labour force, and the reorganization of commercial and mercantile networks surrounding the production process could all generate gains in productivity, even on their own.

Technical change in the eighteenth century is commonly associated with the coming of the factory system, for factories sprang up just as a remarkable wave of technical innovation spread across the cotton industry. A sharp historical contrast was drawn thereafter between two models of industrial organization – one associated with innovation, machinery and the factory, the other with backwardness, hand techniques and the domestic system. But real life, as it always does, differed from the models. There was in fact a great range of forms of industrial organization: the factory and the domestic system lived side by side, even within the same industry. And to say 'factory' was by no means to say 'progressive' or 'machinery'. For the factory existed long before and independent of the machinery. There were jenny shops which brought together spinners, there were handloom-weaving sheds, glass and paper factories, early machine works and, from the early years of the eighteenth century, there were foundries.

Many of the factories established in the late eighteenth century in the textile and other industries did indeed exhibit many progressive features. With hindsight we can find in them exercises in the division of labour, standardization of production, order, job specification and the employment of as much unskilled labour as possible. There were works like Boulton and Watt's Soho Foundry, the machine shops of Bramah and Maudslay, and the remarkable prototype of assembly-line production to be seen in Samuel Bentham's manufacture of ships' biscuits for the navy.

Yet if aspects of the early factory system seemed so modern, in many other ways this was a very traditional form of industrial organization. There had been centralized production units for at least two centuries before the first great cotton-spinning mills. The 'houses of industry' which formed part of the poor law administration bore a closer relation to the first factories than most historians have cared to admit. And many early factories, particularly the country mills, were run on the example of an army barracks, using contract labour, and imposing paternalist, feudal regulations. Furthermore, these factories were rarely seen as an alternative to the domestic system, but only as an adjunct to it. For they were nearly always complemented by widespread putting-out networks, sometimes even within the same process (as for a time in both spinning and weaving), and certainly between processes. In addition most textile factories were run on the basis of family organization.

Many so-called factories were, in addition, more like a collection of separate workshops than a centralized unit. In the cutlery trades, metalworking and the potteries, the factory was only a building where skilled artisans carried on their separate crafts. This was even the case in the great Soho Foundry where Boulton and Watt had set out to establish an orderly flow of work and to impose a strict disciplinary code on their workforce.

For the foundry was essentially an amalgam of workshops, and the workforce was made up of skilled craftsmen employed on piecework.

Neither can it be assumed that the domestic system was outmoded and confined to the hinterland of the factory by the beginning of the nineteenth century. For manufacture continued under domestic and workshop forms of production in most assembling industries, metalworking and even textiles, including cotton. As late as 1851 there were still tens of thousands of cotton handloom weavers at work in their houses.[31]

In fact, there were many advantages to the domestic system. Capital could easily be shifted from one industry to another, for little of it was tied down in heavy fixed investment in building and equipment. It was indeed frequently transferred completely in and out of industry and more traditional investment outlets in land and the food and drink trades.[32] Industries organized on small-scale or domestic lines also drew on a flexible labour force, one which seemed to be both infinitely expandable and easy to cast off. For their labour force was originally a part-time one supplementing a meagre family income earned in agriculture, mining or fishing. As a result, such early industrial labour was on the whole not well organized to enforce or even to claim high wages. Most worked for less than subsistence wages.

In spite of such advantages to the domestic system, the pundits and leading entrepreneurs of the period still believed the factory would allow them to use a more disciplined and even cheaper unskilled labour force. They sought to displace skilled labour through factory organization and mechanization. But the introduction of factories and machinery was also dependent on an existing and growing foundation of metalworking, engineering and metal-extracting skills to build the machines and the factories, and to devise, set up and run their systems of power.

Not only did factories and machinery create such demands for more skilled labour, they also eventually drew in large reserves of basic, less skilled labour as they expanded on the foundation of the existing technology. The factory system and the domestic system both spread in the eighteenth and the nineteenth centuries. Domestic labour complemented factory labour over the whole period. The existence of a large domestic labour force cut the costs of full mechanization and offset the effects of cyclical fluctuations.

While highly organized groups of skilled artisans in both town and country areas fought the displacement of their skills by new machinery, large groups of underemployed domestic workers helped to shore up the new technology by reducing risks and costs. The struggle between artisan and machinery, which started in the eighteenth century, had, however, by the nineteenth extended to a wider struggle between these hand and mechanized branches of the textile industry. This struggle between artisan and factory was also transmitted to the inside of the factory itself as a struggle between skills and machines. By the 1830s Andrew Ure could represent the factory system and the machine as the method of disciplining labour and dispensing with difficult groups of skilled workers. But the most difficult group of workers, the mulespinners, managed to maintain their position in face of the threat from the self-acting mule, by fighting for control over the new machinery, rather than by trying to stop its progress altogether. Inside the factory and out, the relation of workers to those changes in technology and organization which allowed for great gains in industrial output during the Industrial Revolution was reflected in their day-by-day workplace struggles.

The economic impact of technical change

Improvements in industrial organization within the domestic system or inside the factory could affect production as much as could technical change itself. The effects of new technologies on productivity were also frequently hidden by other changes in factor inputs, that is, labour and capital. But the transformation which took place during the eighteenth and early nineteenth centuries in technologies and in organization, as well as in capital investment and employment, can be judged by the remarkable shifts we see in a number of basic economic indicators

– industrial structure, distribution of employment, and distribution of national income between sectors.

If we look at the general trend of change in industrial structure in the eighteenth and nineteenth centuries in Britain, France, Germany and the United States, the most striking change was the reduction in the share of agriculture. In most industrialized countries the share of agriculture fell from over half of total production in the premodern period to 20 per cent or less in this century.

Table 11. *Share of national product (percentages)*[33]

		Agriculture	Industry	Services
England and Wales	1688	40	21	39
	1770	45	24	31
Great Britain	1801	32	23	45
	1841	22	34	44
	1901	6	40	54
France	1789–1815	50	20	30
	1825–35	50	25	25
	1872–82	42	30	28
Germany	1860–9	32	24	44
	1905–14	18	39	43
USA	1839	69	31	—
	1879	49	51	—

In the case of Britain, it is notable that in spite of increases in output in the eighteenth century, there was then no swing away from agriculture. The first major reduction in its share of total output did not come until the early nineteenth century. For most of the eighteenth in fact, the structure of national product did not change a great deal. Some change did seem to be under way by the end of that century, but the effect of the Napoleonic Wars was to cancel this out. A definite trend away from agriculture and in favour of manufacture became notable only from the 1820s. The structure of Britain's national output in more detail looked like this:

Table 12. *Structure of national product, 1688–1841 (percentage share of national income)*[34]

	1688	1770	1801	1811	1821	1831	1841
Agriculture	40	45	32.5	35.7	26.1	23.4	22.1
Manufacture, mining and building	21	24	23.6	20.8	31.9	34.4	34.4
Commerce/trade and transport	12	13	17.5	16.6	15.9	17.3	18.4
Professional and domestic service	15	11	11.3	10.4	5.7	5.7	6.0
Government and defence	7	4	9.8	10.8	13.1	11.6	9.6
Housing	5	3	5.3	5.7	6.2	6.5	8.2

The predominance of agriculture in the distribution of national income until the early years of the nineteenth century was complemented by the growing importance of manufacturing. For manufacturing industries raised their share from one fifth to one quarter of national product, then lost ground again in the first years of the nineteenth century, only to forge ahead to more than a 30 per cent share in the 1820s.

The industrial distribution of the labour force shows some similar trends. Though we do not have sufficient data to detail the course of change in the eighteenth century, there was certainly a notable change in the positions of agriculture and manufacturing in the first twenty years of the nineteenth century.

Though the share of the labour force in agriculture showed a marked decline over the first half of the nineteenth century, the total numbers of its workers did not reach its peak until the mid-nineteenth century. In 1851, with more than 2 million employees, it was still the most important British industry. Nevertheless, in the first years of the nineteenth century a main shift of labour took place towards manufacturing, mining and building. Manufacturing expanded in the period just after the Napoleonic Wars so that by 1841 it claimed one third of the labour force.[35]

Table 13. *Distribution of the British labour force, 1801–1901 (percentages)*[36]

	1801	1821	1841	1861	1881	1901
Agriculture, forestry and fishing	35.9	28.4	22.2	18.7	12.6	8.7
Manufacture and mining	29.7	38.4	40.5	43.6	43.5	46.3
Trade and transport	11.2	12.1	14.2	16.6	21.3	21.4
Domestic and personal	11.5	12.7	14.5	14.3	15.4	14.1
Public and professional	11.8	8.5	8.5	6.9	7.3	9.6

When agriculture's share of both material income and the labour force did decline, this was no reflection on its economic performance. It did not form a 'declining sector' beside the more progressive manufacturing industries. Its relative 'decline' tells us as much about the rising productivity of agriculture as it does about the growth of industry. For until 1870 it was higher productivity in agriculture combined with a low income elasticity of demand for its products which accounts for much of the decline in its share of the total labour force. After 1870 food imports play a much greater part in the redistribution of the labour force away from agriculture.

The strength of agriculture as against the growing but definitely uncertain place of manufacturing was apparent in the economic indicators until the second quarter of the nineteenth century: the rapid rate of technological change in the years traditionally identified with the Industrial Revolution seems to have had little macroeconomic impact at the time. It affected productivity and output in individual industries, establishing above all a trend of improvement in each sector. But the experience in the eighteenth century was of a secret but gathering force. And the change it made to the economy as a whole, though delayed, was ultimately marked and abrupt. The technological spurt and organizational changes of the eighteenth century did not make their mark on the wider economy until the decades of the 1820s to the 1840s.

These aggregate economic indicators tell us nothing, however, about the impact of industrial growth and technical change on wages and employment. And indeed, there are no aggregate measures of this. Yet economic historians have, in general, assumed a favourable impact on employment – that machinery did not substitute for labour, and may in some cases have intensified it.[37] This is questionable, and is anyway something we cannot know in the aggregate. The impact of this growth and change on wages is at least as difficult to gauge, though economic historians are keen to provide the numbers, covered though they are with the thorns of the standard of living debate. It is interesting to note, however, that recent estimates of the standard of living by Lindert and Williamson favour the optimists' case, but show virtually no trend of improvement in earnings before the 1820s. This is in spite of the upward bias of their estimates of industrial wages, calculated only on the basis of male workers in the highest paid sectors. If, as this chapter has argued, industrial growth was gradual but substantial over the whole of the eighteenth century, and across many industries, we must be left to ask who benefited. It seems, on present estimates, unlikely to have been the workers.[38]

CHAPTER 2

Political Economy and the Growth of Manufacture

A wide range of industries experienced improvements in productivity and increases in output over the course of the long eighteenth century. The dramatic change in a very few industries at the end of that period should not overshadow the breadth of real improvement from the earliest years of the century. If historians have failed to give due weight to this earlier and more widespread economic growth, they have been deaf to the voices of contemporaries. If from the usual evidence for economic change found in the data on output we turn to the opinions of contemporaries, industrial growth and change appear both obvious and of intrinsic interest. Daniel Defoe's *Tour through the Whole Island of Great Britain* confirmed this even as early as 1720:

> New discoveries in metals, mines and minerals, new undertakings in trade, engines, manufactures in a nation pushing and improving as we are; these things open new things every day, and make England especially shew a new and differing face in many places, on every occasion of surveying it.[1]

Defoe's views find their context in the writings of many other economic commentators of the time and act as a sounding of opinion. In this world of opinion and observation we find a way of moving beyond the skeletal suggestions of numerical indices. For contemporaries offered their own colourful accounts and analyses of the course and structure of eighteenth-century industrial growth. Two problems arise, however. First, which contemporaries do we heed, and which do we dismiss or simply not include in our opinion survey? And secondly, how do we assess the evidence of opinion as against the real course of events? A random survey of any person's views is obviously of little use, and one does better to confine research to those who had a direct interest in the economy of the day. We will therefore look only at the economic commentators and search out their views of the role and development of manufacturing industry in eighteenth-century Britain.

Even in confining ourselves to economic thinkers, we still meet many historians who find such contemporary opinion analytically backward or simply ignorant of the real economy. But such historians frequently take little care in understanding their economic thinkers, and in particular the interplay between analysis and description in their writings. Those who did not give much of space to descriptions of actual machines and factories could well be saying a great deal about the contemporary role of manufacturing industry, but they might make their case not by description but by economic analysis. Such economic analysis might take the form of models, the formulation of broad principles of economic growth and decline, and the interplay between theory and policy debate.[2]

Early eighteenth-century commentary

'Manufacture' was a subject of considerable interest in the seventeenth century, and this interest continued into the eighteenth, but the context for discussion changed. Recent historians have praised seventeenth-century writers, but they have neglected or dismissed those who more immediately preceded Adam Smith. But for all this, the economists before Adam Smith had a real sense of the quickening of the economy in preparation for its industrial spurt. They discussed the rapid growth of home markets, the effects of new markets in the fast-growing American economy, and the emergence of new and vigorous industries in the town and country areas of Lancashire, Yorkshire and the Midlands.

These early economists furthermore related their discussion of manufacture to wider economic issues, including the growth of output, the relation of town to country, and of industry to agriculture, the condition of labour, capital accumulation and the relation of organizational and technical improvement to social change. Defoe hailed the period since 1680 as 'a projecting age when men set their heads to designing Engines and Mechanical Motion', and other eighteenth-century writers pointed out that it was higher consumption and the creation of new demands which provided the greatest incentive to efficiency, industry and invention.[3]

One of the earliest and indeed most remarkable of these commentators was Henry Martyn who wrote 'Considerations on the East India Trade' (1701). J. R. McCulloch, the nineteenth-century economist and collector of economic tracts, had a high regard for Martyn and commended him especially for his analysis of the division of labour.[4] Martyn, like Adam Smith long after him, connected the division of labour to the market, on the one hand, and to technical change on the other. He first argued that the clearest route to manufacturing development and higher productivity lay in getting rid of trade strategies and technologies which used more rather than less labour. Policies which aimed at full employment by using more labour at the workplace than was strictly necessary were a delusion. 'Make work' policies 'reduced the business of the people by making our manufactures too dear for foreign markets'.

> If the same work is done by one which was before done by three; if the other two are forc'd to sit still, the kingdom got nothing before by the labour of the two, and therefore loses nothing by their sitting still.[5]

Martyn thought that those left unemployed through the use of more efficient technologies would be much better used in less skilled trades which produced more standardized commodities. He argued that there should be a division of labour between more and less skilled trades, and within trades between more and less skilled processes. More world trade and wider markets including the East India trade would provide cheaper consumer goods. They would also provide incentives for higher productivity at home by leading to 'the invention of Arts, Mills and

Engines to save the labour of hands in other manufactures'.[6]
Trade and technical change were closely interrelated; more
trade would rationalize the division of labour and especially the
division of skills among trades.

Martyn thought the East India trade would introduce 'more
artists, more order and regularity' in English manufactures, in
other words a division of labour, and he illustrated this principle
with detailed descriptions of the processes of manufacture in
textiles, watchmaking and shipbuilding.[7] Martyn not only argued
that more world trade would positively promote technical
change; he pointed out that new attitudes to technology should
be fostered in England. He drew attention to the large numbers
of mills and engines in Holland:

> But has more than only one sawmill been seen in
> England? . . . by a wonderful policy the people here must not
> be deprived of their labour; rather every work must be done
> by more hands than are necessary.[8]

This interest in technical change, division of labour and
markets continued to fascinate contemporaries throughout the
eighteenth century in spite of the changing economic climate of
harvests, prices, labour supply and wages.[9] The works of John
Cary, the late seventeenth-century Bristol sugar merchant, were
reprinted in 1715 and 1745. Against the general tenor of
opinion in favour of lower wages and schemes for employing the
poor, Cary presented very detailed work on wages and productivity, new manufactures and technical change. He pointed
out the extent to which technical change had succeeded in
reducing costs in a whole series of industries including sugar
refining, distilling, tobacco manufacture, woodworking and lead
smelting.

> There is a cunning crept into the trades – the clockmaker
> hath improved his art to such a degree that labour and
> materials are the least part the buyer pays for. The variety of
> our woollen manufacture is so pretty, that fashion makes a
> thing worth twice the price it is sold for after . . . artificers, by

tools and lathes fitted for different purposes, make such
things as would puzzle a stander by to set a price on,
according to the worth of men's labour . . . new projections
are every day set on foot to render the making our woollen
manufactures easy, which should be rendered cheaper by the
contrivance of manufacturers not by the falling price of
labour; cheapness creates expense, and expense gives fresh
employments, whereby the poor will be still kept at work.[10]

Joshua Gee wrote in 1729 of the technical advance in the
copper and brass industries, and of the emergence of the new
hardware, steel and toy trade. He was particularly fascinated by
the Italian silk-throwing machine 'which with a few hands to
attend it will do more work than an hundred persons can do at
throwing by our method'.[11]

If machinery and new methods were noticed, so too were new
ways of organizing industry in order to take advantage of the
benefits Henry Martyn ascribed to the division of labour. There
was perhaps no better description of this than Daniel Defoe's
observations on the highly sophisticated domestic system of the
West Riding of Yorkshire. Here was a countryside which
seemed 'one continuous village'. To every considerable house
was attached a 'manufactory or workhouse'; each had its own
stream of running water and easy access to coal fuel, and each
kept a horse or two and a cow or two with enough land to feed
them. Amongst the manufacturers' houses were 'scattered an
infinite number of cottages or small dwellings, in which dwell
the workmen which are employed, the women and children of
whom are always busy carding and spinning'. The workmen
were all employed in the clothiers' manufactories, 'a houseful of
lusty fellows, some at the dye-fat, some dressing the cloths,
some at the loom, some one thing, some another, all hard at
work and full employed upon the manufacture, and all seeming
to have sufficient business'.[12]

What is especially interesting about Defoe's description is
that it was not about our usual image of domestic industry. It did
not point out the advantages of rural versus urban industry, and
it did not dwell on the special features of family-based or

peasant industry. It was actually a description of a workforce dwelling in the countryside, and what struck Defoe was the division of labour in a rural area between argiculture and industry. For Defoe saw few people out of doors in the area, and little corn. The inhabitants imported their corn from Lincolnshire, Nottinghamshire and the East Riding, and the clothiers bought their beef in the market in Halifax. He also described a division of labour within the workshops. This was not family production in peasant households, but the employment of workers in assigned tasks and, as a result, it was a 'populous and wealthy region'.[13]

As Dorothy George argued many years ago, this rural manufacture of the West Riding was described in such detail by Defoe precisely because it was very different from the organization of the cloth manufacture in the South. The West Riding contained a dispersed rural industry, but it was not, as Defoe noted, an ideal type of the domestic system, for the larger houses were little manufactories owned by master clothiers who employed journeymen and apprentices on the side and in their own cottages. Those cottages most frequently had no landholding attached to them. Arthur Young confirmed this when he surveyed the area in 1795: 'Their land is generally at 40s. an acre, it is only the master clothier that has it; the loom men in the cottages have none.'[14] The particular area which Defoe described was also in the course of turning over from woollen production to worsted manufacture, and this was an industry which from the very beginning was in the hands of those employing a large capital and labour.

Still, the West Riding impressed not only Defoe but also Josiah Tucker a few decades later for the absence of anything like the great gulf between the wealthy clothiers and the poor weavers and spinners which they saw in the South. Tucker found that in Gloucestershire, Wiltshire and Somerset the manufacture was carried on under quite different conditions, and the effects were accordingly that

> one person with a great stock and large credit, buys the wool, pays for the spinning, weaving, milling, dyeing, shearing,

Manufacture and the Economy

dressing, etc. That is, he is the master of the whole manufacture from first to last and perhaps employs a thousand persons under him. This is the clothier whom all the rest are to look upon as their paymaster. But will they not also sometimes look upon him as their tyrant?[15]

Defoe described the growth of industrial regions in Lancashire, Yorkshire and the Midlands, and pointed out the extent to which these regions were becoming known for their specialized manufactures, manufactures which as a result were subject to all the gains in productivity to be had through the division of labour and technical change.

Many of the industries cited in early to mid-eighteenth-century economic commentary had indeed already made a significant organizational transition away from rural by-employments to specialized town or country industries. Joan Thirsk has argued only recently that the seventeenth-century connection between dairy farming and stocking knitting was broken by the eighteenth. Knitters now lived increasingly in towns with little part-time interest in farming. There were many who 'never tasted the independence of the hand knitters. From the outset they were enslaved, like the spinners and weavers of cloth, working for a clothier for small wages, so that they lived very poorly.'[16] But contemporaries such as Dean Tucker were equally aware of the divide between employer and employed: the weavers of the West of England cloth industry were employees who had lost their connection with the land. Most owned no more than a garden, some owned houses in the towns, and were rentpayers and the tenants of the clothiers.[17]

The middle years

Specialized manufacture, be it in town or country, was what met the eye of the economic commentator of the early eighteenth century. This theme continued to fascinate writers into the middle of the century, but commentators now also turned their thoughts to some of the industry's social implications. Josiah

Tucker, Malachy Postlethwayt, and the author of *Reflections on Various Subjects Relating to Arts and Commerce* . . .[18] considered ways of introducing new industries, the prospects and possible dangers of introducing new techniques to displace labour, and the best locations for particular industries. They debated the benefits of encouraging the immigration of foreign artisans as a means of reducing the wages and increasing the discipline of English labour,[19] or as a source for new industries and skills.

An earlier fascination with labour-saving technical change became a much more complex consideration as the century progressed. Against a backdrop of a steadily rising population, the pros and cons of labour-saving inventions were carefully considered. The author of *Reflections on Various Subjects* (1752), for instance, wrote that the machines 'did the work truer and better than the hand', and the labour saved by them was so great 'that they who use the machine must undersell the others in a vast disproportion'.[20] Still he did not consider it easy to determine the pace at which technical change ought to be allowed to proceed. He finally decided that 'engines' might be introduced with no problem, in the first case where they did jobs that could not be done at all by hand, as with pumps, fire engines, looms, wine and oil presses; and secondly where the commodities concerned could not be produced at all except by machine, as in papermaking and iron-processing machinery, and fulling mills. Another consideration was the type of economy – was this a country with a large sector of foreign trade, or a fairly isolated community? Commercial states that had to produce cheaply to gain foreign markets had no option but to use labour-saving techniques. But those with little trade, where the technological unemployment created might adversely affect home markets, did have some justification for holding back or preventing the use of machinery.[21]

Postlethwayt rejected such arguments, limiting his reservations to the use of machinery in agriculture. He thought that the skill of workmen would lead naturally to invention, invention which would not, contrary to popular opinion, reduce employment. It would lead, instead, to more employment 'by multiplying works and increasing the produce of the balance, which never ceases to increase home consumption'.

We do not see any objection that can be made to the economizing of time, or facilitating the work of manufactures which may not be equally well made to all inventions of new fashions, or of new stuffs, but which the old are forgot . . . I believe no man will say it is the interest of a nation to prohibit new manufactures, in order to favour the workmen employed in the old.[22]

But still, Postlethwayt believed with the author of *Reflections on Various Subjects* that home markets had to be maintained in order to prevent any English industry from being undermined by foreign imports. The best security for this market was in the 'cultivators of the soil', and 'every machine tending to diminish their employment would really be destructive of the strength of society, of the mass of men, and of home consumption'.[23]

The complications of this intellectual reaction to new technology reflected the frequently opaque world of the development and introduction of new technologies in different industries and regions. The other side of the story of transitions in economic organization and technology was always shaped by the reasons behind the resistance to, or simply failure to take up, such developments. This is an important subject, and one which comes up in more depth later in the book. But it is striking that, in spite of a definite awareness of the social impact of such improvement, there does seem an overriding interest among these writers in the potential markets, skills, ingenuity and suitable price of labour, and the possibilities of saving labour through technical change.

By the 1760s and 1770s a work like Anderson's *Historical . . . Deduction of the Origin of Commerce* could be organized around cataloguing new manufacturing industries, and describing the new machinery daily being introduced into particular trades. And William Kenrick unreservedly assumed that any well-governed nation would expedite the introduction of labour-saving machinery, as the best means of gaining foreign markets.[24] By this time, too, writers on political economy were concerned with problems of labour discipline and looked for alternatives to the present system, alternatives not yet perceived in machine technology, but certainly seen in some form of factory organization. Where Postlethwayt had claimed that the poor were industrious and deserved good wages, J. Cunningham in 1770 objected that the so-called industry of the poor was only predicated on a series of Elizabethan statutes to enforce labour and regulate its price. But this had clearly proved insufficient, for

> the lower sort of people in England from a romantic notion of liberty, generally reject and oppose everything that is forced upon them; and though, from a fear of punishment, you may oblige persons to labour certain hours for certain wages, you cannot oblige them to do their work properly. If they work against their wills, they may slight their work, and our foreign trade may be hurt.[25]

The answer to the problem might be found in the type of factory discovered by him in Abbeville. Six hundred workers came to work and left it at the beat of a drum, and 'each branch had a distinct foreman who disciplined the workmen so as to make them excel in every branch of the whole'.[26]

The stage was thus set by the later eighteenth century for Adam Smith's analysis of the significant connections between the expansion of markets, the division of labour and technical change, acting in concert to turn the engine of economic progress. The eighteenth century was not marked by a gap between the insights of the seventeenth-century mercantilists and the advances of Adam Smith, but by continuous analysis from the later seventeenth century of the connections between markets, technical change and industrial expansion.

Adam Smith

It was on this edifice of longstanding interest throughout the eighteenth century in the division of labour and technical change that Adam Smith developed the division of labour into a principle underlying the whole mechanism of the economic and political institutions he was analysing in the *Wealth of Nations*. Smith's discussion of manufacturing industry in the eighteenth century is frequently misunderstood by economic historians who fail to find many direct references in his work to cotton, steam, iron and factories. But Smith formulated distinctive principles of manufacture within the framework of a model of economic growth and development. Within the terms of this model he explained the emergence and performance of different forms of organization of industry. Like his predecessors, he praised the achievement and technical aptitude of independent skilled artisans, but he was considerably less sanguine about the gains and prospects of rural domestic by-employments.

Indeed, we find in Smith the beginnings of a debate about rural industry. From the outset he argued that it was the division of labour or specialization of economic activities which generated gains in productivity, and this division of labour was hinged in turn on the development of the market and capital accumulation. The size of the market determined the extent to which any trade could be carried on as a separate full-time employment and thus affect increases in productivity: 'It is found that society must be pretty far advanced before the different trades can all find subsistence . . .'[27]

> There are some sorts of industry, even of the lowest kind, which can be carried on nowhere but in a great town. A porter, for example, can find employment and subsistence in no other place. . . . In the lone houses and very small villages of the Highlands of Scotland, every farmer must be butcher, baker, brewer for his own family.[28]

The labourer who had to take on multiple employments because the market was not large enough to sustain any single occupations could not increase his dexterity, save his time, or apply himself to technical improvements. There were, therefore, strict limitations to his potential productivity.

The extent of specialization was also determined, as Smith continued, by the size and rate of increase of the capital stock. An employer's capital had to be sufficient to employ a particular labourer at any single occupation. Any increase in a capital stock would also tend to raise wages, which in turn created incentives for a division of labour and higher productivity.

> The owner of the stock which employs a great number of labourers necessarily endeavours, for his own advantage, to make such a proper division and distribution of employment, that they may be enabled to produce the greatest quantity of work possible.[29]

The historical model and the artisan

In Book III Smith demonstrated how this framework – division of labour, market and capital – came together in a dynamic model of the development of agriculture and industry, town and country. The model and the historical economics of this Book form the reference point for views of manufacture expressed by Smith elsewhere in the *Wealth of Nations*. Here Smith argued that there was a 'natural progression' of economic development. The 'natural' (which was not necessarily the actual) course of development was a model of balanced economic growth based in the first instance upon agriculture.

On the basis of these principles Smith then constructed an historical model of the development of manufacturing industry in relation to agriculture and of the development of towns in relation to the countryside. He argued that there was a model or 'natural' progress of economic development from agriculture to manufacture and thence to foreign commerce. 'Manufactures for distant sale' might 'grow up of their own accord, by the gradual refinement of those household and coarser manu-factures which must at all times be carried on in even the poorest and rudest countries'. Based on domestic raw materials, they generally sprang up in an inland country which produced an agricultural surplus that it, in turn, found difficult to trade, owing to high transport costs. The surplus, however, made basic needs very inexpensive, encouraging the immigration of a larger labour force. These workmen

work up the materials of manufacture which the land produces ... they give a new value to the surplus part of raw
produce ... and they furnish the cultivators with something
in exchange for it that is either useful or agreeable to
them.... They are thus both encouraged and enabled to
increase this surplus produce by a further improvement and
better cultivation of the land; and as the fertility of the land
had given birth to the manufacture, so the progress of the
manufacture re-acts upon the land and increases still further
its fertility.[30]

This was the 'natural progress' from agriculture to manufacture
and thence to foreign commerce, praised by Smith for leading
both to the most rapid rates of growth. Such 'natural progress',
Smith conceded, had actually taken place in certain parts of
England, where some cities had arisen on the basis of rural
industries complementing regional agricultural surpluses.

In this manner have grown up naturally, and as it were of
their own accord, the manufactures of Leeds, Halifax,
Sheffield, Birmingham and Wolverhampton. Such manufactures are the offspring of agriculture.[31]

This complementary development of agriculture and manufacturing created, Smith argued, the best conditions for working
men and women, for it guaranteed their 'independence'.
Agricultural development which made for cheap food allowed
workers to leave their employers and to become independent
labourers and artisans. Where agriculture was fully developed
and surpluses and provisions were plentiful, labourers could
'trust their subsistence to what they can make by their own
industry'.

Nothing can be more absurd, however, than to imagine that
men in general should work less when they work for themselves, than when they work for other people. A poor independent workman will generally be more industrious than
even a journeyman who works by the piece. The one enjoys

the whole produce of his industry; the other shares it with his master . . . the superiority of the independent workman over those servants who are hired by the month or by the year, and whose wages and maintenance are the same whether they do much or do little, is likely to be still greater.[32]

Accordingly, most landlords, farmers and masters preferred circumstances of poor harvests, low agricultural output and high food prices, for in these dear years they 'make better bargains with their servants . . . and find them more humble and dependent'.[33]

It seems at first surprising that Smith could have considered the division of labour and specialization consistent with the prospect of growing numbers of independent artisan labourers. He had first acknowledged the existence of the artisan labourer early on in his chapter of wages:

It sometimes happens, indeed, that a single independent workman has stock sufficient both to purchase the materials of his work, and to maintain himself till it be completed. He is both master and workman, and enjoys the whole produce of his own labour, or the whole value which it adds to the materials upon which it is bestowed. It includes what are usually two distinct revenues, belonging to two distinct persons, the profits of stock, and the wages of labour.[34]

He also added that such cases were not very frequent. But his view was that in the progress of commercial society, wage labour would not be universal: to an increasing extent the independent artisan also enjoyed his place. He might also contribute to increasing productivity, for his existence was by no means inconsistent with the technical division of labour. The workman who produced but one type of file or one part of a watch could well be an independent artisan; he was the archetypal workman-inventor who developed and perfected the tools, machines and skills of a trade. The progress of opulence might engender the artisan, but the artisan in turn would contribute to the division of labour on which that progress ultimately depended.

The critique of the putting-out system

Yet Smith's 'natural progression' was not an historical model of European economic development. With regret, he traced how European policies and follies had generally resulted in an opposite course of development – not from agriculture to industry and commerce, but from foreign commerce and industry to agriculture. Smith's belief in a form of economic development which held out a future to independent artisanal manufacturing was therefore at odds with what had actually happened in the European economies. Manufacturing had not in general arisen from the 'schemes' and 'projects' of individual merchants setting up luxury manufactures. The merchants and artificers, 'in pursuit of their own pedlar principle of turning a penny wherever a penny was to be got', exploited the countryside and turned the terms of trade in favour of the town. If the towns thus became in the end the cause of the improvement of the country, this development had been slow and uncertain.

> Compare the slow progress of those European countries of which the wealth depends very much upon their commerce and manufacture, with the rapid advances of our North American colonies, of which the wealth is founded altogether in agriculture.[35]

Finally, the capital created by a merchant, 'who was not necessarily the citizen of any particular country', was an unstable possession until part of it was reinvested in the land. An economy with a significant agrarian base was much more likely to have a strong and stable political and social structure.

Adam Smith then analysed different types of nonfactory production. He assessed the merits of independent proprietorship as opposed to the putting-out system, and argued that the industrial and agrarian history of various regions lay behind the many problems entailed in the domestic form of industry. Later economists such as John Stuart Mill and recent historians such as Joan Thirsk have found in Smith's account a 'grotesque caricature of the weaver-farmer' and a 'simplified, partial and occasionally harsh view of the domestic system'.[36] But the alternative Smith had in mind was not the factory slave, but the highly paid wage labourer or independent artisan. The domestic system he wrote of was exploitative and poor, not specialized, innovative and highly paid.

The rural industries and putting-out system described by Smith were undercapitalized and underpaid. It was because of inadequate capital to provide for full-time occupations that the underemployed country weaver 'sauntered a little in turning his hand from one employment to another'. Such underemployed workers were, furthermore, willing to work at other trades for less than the customary rates. These rural industrial by-employments were seen at their worst in the Highlands of Scotland; the monuments of rural poverty, not of agricultural wealth. Many of the manufactures Smith described were not those which had grown 'naturally' out of agriculture (as in 'Leeds, Halifax, Birmingham and Wolverhampton'), but the 'unnatural' extensions of commerce, monopolistic restriction and mercantile greed. There was perhaps no better example of this than the linen industry of his own country. There, wealthy merchants had tried to achieve the repeal of the linen bounty for their own individual gain. Smith pointed out that most of the labour of producing linen cloth went into the preparation and spinning of the yarn, and 'our spinners are poor people, women commonly, scattered about in all the different parts of the country, without support or protection'. But the great merchants and master manufacturers wanted to sell the final cloth as expensively, and to buy the yarn as cheaply as possible. 'It is the industry which is carried on for the benefit of the rich and the powerful, that is principally encouraged by our mercantile system. That which is carried on for the benefit of the poor and the indigent, is too often either neglected or oppressed.'[37] The countryside and its workforce had been put at a disadvantage by a long history of economic policies designed to promote the interests of urban incorporated industries at the expense of agriculture and other rural enterprises. And urban artisans had falsely credited themselves with superior skills, established and buttressed by resort to monopolies and corporate restrictions.

Smith's few asides on the poverty of those domestic industries practised as by-employments did not constitute an extended analysis of this industrial organization. If we are to understand his perspectives we must look to his much broader enquiry into the division of labour. It is important to note that Smith and his contemporaries were writing about the prospects and problems of small-scale industry, not mechanized factories. Such large-scale works as existed at the time were remarkable, but very few and far between. Most of those that did exist depended on the special market of government and military contracts, including Crawley's ironworks, the Chatham dockyard, the Carron Works, Walker's and Wilkinson's ironworks. Smith's criticisms of small-scale domestic industry were not directed at the size of this industry, but at its organization. Those small-scale industries organized on the basis of extensive putting-out networks and set up by wealthy mercantile interests frequently exploited the countryside and especially the country labourer. But domestic industries which had arisen on the basis of local initiative and local agricultural surpluses had many benefits, especially for those practising them. Rural manufactures might coexist along with agriculture, but only in such a way as to guarantee the independence of artisans and labourers.

The end of the century: critiques of the division of labour

A widespread debate on domestic industry and putting-out systems was taken up by Smith's contemporaries and followers, but it was one which singularly failed to make a distinction of ownership and control over production and its output. Most writers in fact focused on only two problems: the economic impact of manufacturing on agricultural improvement, and the social implications of the division of labour.

The eighteenth-century rural industries did not impress either James Anderson or Arthur Young, for they both complained of their effects on local agriculture and they wrote of these industries from the viewpoint of the landlord or wealthy tenant farmer. Anderson thought rural manufactures lured labour away from agriculture and encouraged landlords to break up their tenancies into small plots to rent to cottagers instead of to tenant farmers. This would result in an unstable social order and lower agricultural productivity. Industry was preferably organized away from agriculture and centralized so as 'to be carried on by people in concert' who would all 'work in one place'.[38] Anderson's worries over the effects of domestic industry on agriculture also loomed large in Arthur Young's work. In his controversy with Mirabeau between 1788 and 1792, in the *Travels in France* and the *Tour of Ireland* Young pursued the issue entirely with regard to the productivity of labour in agriculture. He found those provinces known for their manufacturing – Normandy, Brittany, Picardy, and the Lyonnais – to be 'among the worst cultivated in France'. 'The immense fabrics of Abbeville and Amiens have not caused the enclosure of a single field.' 'The agriculture of Champagne is miserable, even to a proverb: I saw there great and flourishing manufactures, and cultivation in ruins around them.'[39] Examples drawn from Britain and Ireland confirmed him in his view that poor cultivation was to be attributed entirely to 'manufacture spreading into the country, instead of being confined to the towns'.[40]

The other side of the coin on domestic industry was Dugald Stewart's defence of the 'sauntering weaver'.

> Though it follows that a domestic manufacture must always be a most unprofitable employment for an individual who depends chiefly for his subsistence on the produce of a farm, the converse of the proposition requires some limitations. A man who exercises a trade which occupies him from day to day must of necessity be disqualified for the management of such agricultural concerns as require a constant and undivided attention . . . but it does not appear equally evident how the improvement of the country should be injured by his possessing a few acres as an employment for his hours of recreation. . . .

'Occasional labour in the fields' was better than 'those habits of intemperate dissipation in which all workmen who have no variety of pursuit are prone to indulge'.[41] But, in agreement with Smith, he pointed out that there were rural industries and rural industries: those which produced common rather than luxury commodities were safer regional enterprises.

> The manufacturers of Norwich who deal in fine crape and other delicate stuffs are laid idle three times for every once that the Yorkshire manufacturer who deals in low priced serviceable cloths experiences a similar misfortune.[42]

The debate over domestic industry continued into the first half of the nineteenth century, as economists compared industrial structures across regions and countries, often disputing at cross-purposes over some historical ideal and the worst examples of sweating which pervaded nineteenth-century British industry. J. R. McCulloch, for example, found little in domestic industry to commend itself. Combinations of agriculture and manufacture, he argued, were contrary to the principles of the division of labour: 'I consider the combination of manufacturing and agricultural pursuits to be proof of the barbarism of every country in which it exists.' He derided all schemes to introduce a 'cottage economy', for under such systems it would be impossible to 'realize capital'.[43] Views like these were, however, disregarded by those like J. S. Mill who went back to the example of Adam Smith's 'sauntering weaver', to argue that the division of labour was no necessary criterion of efficiency. Mill argued against Smith:

> This is surely a most exaggerated description of the inefficiency of country labour, where it has any adequate motive to exertion. Many of the higher description of artisans have to perform a great multiplicity of operations with a variety of tools. . . . The habit of passing from one occupation to another may be acquired, like other habits, by early cultivation and when it is acquired there is none of the sauntering which Adam Smith speaks of . . . but the workman comes to each part of his occupation with a freshness and a spirit which he does not retain if he persists in any one part.[44]

Rural alternatives to mechanization and factory industry were actively pursued throughout the early nineteenth century by poor law and other social reformers, by radicals ranging from Cobbett and Owen to the Chartists, and by spokesmen for the handloom weavers. Many of these alternatives were based on a series of historical myths about the characteristics of manufacture before the factory system. These are the myths like those of Frederick Engels –

Before the introduction of machines, spinning and weaving of the raw materials took place in the workers' house. Wife and daughters spun the yarn, which the husband wove or which they sold if the family's father did not process it himself. These weaver families mainly lived in the countryside near the towns, and could do quite well with their wages. . . . In this way the workers vegetated in a rather comfortable existence . . . their material position was far better than that of their successors.[45]

In the course of the transition from the eighteenth to the nineteenth centuries, however, the debate on manufacture and the place within it of rural industry was turned to a new focus on the social and economic aspects of machinery, and to a wider critique of industrialization. Earlier discussion on the characteristics of rural and urban industry became linked by the end of the eighteenth century to a growing concern over the issue of hand techniques versus machinery. Where discussion was based on an assessment of the actual organization of any one industry, as, say, in the 1806 Inquiry into the Woollen Industry, it might well be that there was little to dispute over.

On the whole, Your Committee feel no little satisfaction in bearing their testimony to the merits of the Domestic system of Manufacture; to the facilities it affords to men of steadiness and industry to establish themselves as little Master Manufacturers, and maintain their families in comfort by their own industry and frugality; and to the encouragement which it thus holds out to domestic habits and virtues.

Neither can they omit to notice its favourable tendencies on the health and morals of a large and important class of the community.

   ... nor would it be difficult to prove, that the Factories, to a certain extent at least, and in the present day, seem absolutely necessary to the well-being of the Domestic system; supplying those very particulars wherein the Domestic system must be acknowledged to be inherently defective: for, it is obvious, that the little Master Manufacturers cannot afford, like the man who possesses considerable capital, to try the experiments which are requisite, and incur the risks, and even losses, which almost always occur, in inventing and perfecting new articles of manufacture. . . .[46]

What mattered was not the size of the industrial unit, but the control over the organization and output of that unit. A debate over technology itself which ensued from the end of the eighteenth century followed similar lines in contrasting mechanized techniques with hand and alternative technologies. But contemporaries became increasingly aware, as Smith had been, that the problem was not the technology or even the organization of industry, but its ownership and control.

CHAPTER 3

Models of Manufacture: Primitive Accumulation or Proto-industrialization?

The picture presented by contemporaries of the kinds of economic change foreshadowing industrialization was one dominated by specialization, division of labour, innovation, skill and mechanical contrivance. It was a picture which emphasized the dynamism and spirit of innovation over the whole of the eighteenth century, not just in its last decades. Many contemporaries were well aware, however, of the social and economic problems borne in the wake of certain types of technological and organizational change, especially those associated with the putting-out system. The observations of contemporaries, however, stand curiously at odds with two major current analyses of the phase of transition between the pre-industrial period and the Industrial Revolution. These analyses are, on the one hand, the 'proto-industrial' model, and on the other the Marxist analysis of primitive accumulation and 'manufactures'. Both these models have been very important in recent years in shaping our own ideas and assumptions about the growth, structure and labour force of manufacturing industry before the Industrial Revolution.

There are many problems with both these approaches, and much of this chapter will be devoted to discussing them. But both models are major and stimulating attempts to conceptualize the economic and social structures of the period. Both have provoked considerable discussion and sophisticated research over a large number of countries and regions. Both draw attention to the connections between agrarian change, commercial capitalism and the growth of handicraft production in rural cottages and urban workshops. Whatever their problems, then, we must first understand the framework of analysis they present.

Primitive accumulation and manufacture

The first analysis of industrial expansion before the factory, to which most historians have been tempted to look back, is Marx's model of the phase of manufacture, and with this, his theory of primitive accumulation. Classical Marxist questions concerning the nature and mechanism of primary accumulation, the role of merchant capital, and the advance of the division of labour informed a widespread debate on the nature of the transition to industrial capitalism. Marx defined primitive accumulation as the necessary prehistorical phase of capitalism. It was the process which initially created the capital-labour relation:

> The process which divorces the worker from the ownership of the conditions of his own labour ... a process which operates two transformations ... the social means of subsistence and production are turned into capital, and the immediate producers are turned into wage labourers.[1]

Primitive accumulation was, in the first instance, associated with agrarian change and enclosure movements. The industrial side of the model is, at first, difficult to see. And the agrarian bias of primitive accumulation fitted well with a separate English historical tradition on the decline of the peasantry. Primitive accumulation influenced the tradition of Tawney and the Hammonds, and was also basic to the beliefs of those Oxford founders of economic history – Toynbee and James E. Thorold Rogers, a former Drummond Professor. It is to this historical tradition that we look to find the strongly held belief that the robbery of his just rights from the labourer took the major form of the robbery of his land. Thorold Rogers, on the basis of his massive and much praised *History of Agriculture and Prices*, denounced the practices of landlords and the government from the Middle Ages through to recent times.

> We have been able to trace the process by which the condition of English Labour has been continuously deteriorated by the acts of government. It was first impoverished by base

money. Next it was robbed of its guild capital by the land thieves of Edward's regency. It was next brought in contact with a new more ready set of employers – the sheepmasters who succeeded the monks . . . the agricultural labourer was then further mulcted by enclosures. . . . The poor law professed to find him work, but was so administered that the reduction of his wages to a bare subsistence became an easy process. . . . The freedom of the few was bought by the servitude of the many. . . . Such was the education of which the English workman received from those evil days, when the government employed and developed the means of oppressing and degrading him. It is no moral that he identifies the policy of the landowner, the farmer and the capitalist employer with the machinery by which his lot has been shaped, and his fortunes, in the distribution of national wealth, have been controlled. He may have no knowledge, or a very vague knowledge, as to the process by which so strange, so useful an alternative has been made in his condition. But there exists, and always has existed, a tradition, obscure and uncertain, but deeply seated, that there was a time when his lot was happier, his means more ample, his prospects more cheerful than they have been in modern experience.[2]

But this historical tradition neglected the industrial side of the peasantry's decline. And Marx too wrote less, and less systematically, on the association between the spread of domestic industry and primitive accumulation. Marx certainly recognized the existence of the significant changes taking place in rural manufacturing production. He wrote of the late eighteenth-century and mid-nineteenth-century economic debates on the benefits of domestic industry. He found Mirabeau arguing that 'the isolated individual workshops, for the most part combined with the cultivation of small holdings, are the only free ones'. And he challenged the philanthropic economists and liberal manufacturers – J. S. Mill, Rogers, Goldwin Smith and Fawcett who asked of the landed proprietors, 'Where are our thousands of freeholders gone?' – to go further and to ask

themselves where the independent weavers, spinners and handicraftsmen had gone to?[3] Marx regarded primary accumulation
as the first stage on the way to industrial concentration. He
regarded the private property of the worker in his means of
production as the foundation of small-scale industry, and small-
scale industry as the necessary condition for the development of
social production and the free individuality of the worker
himself. The fragmentation of holdings and dispersal of means
of production entailed in such a system, however, made it
'compatible only with a society moving within narrow limits'.

> Its annihilation, the transformation of the individualistic and
> scattered means of production into socially concentrated
> means of production, the transformation of the dwarf-like
> property of the many into the giant property of the few . . .
> this terrible and adversely accomplished expropriation of the
> mass of the people forms the pre-history of capital . . . private
> property which is personally earned . . . is supplanted by
> capitalist private property, which rests on the exploitation of
> alien but formally free labour.[4]

He argued that the destruction of the subsidiary trades of the
countryside went hand in hand with the expropriation of a
previously self-supporting peasantry. Only through the destruction of rural domestic industry could an adequate home
market be provided for the capitalist mode of production. Yet he
also pointed out that such domestic crafts did not simply dis-
appear before the emergence of large-scale industry. This
'manufacturing period' did not 'carry out this transformation
radically'. It always rested on the 'handicrafts of towns', and the
domestic subsidiary industries of rural districts, but it destroyed
these in one form and resurrected them again elsewhere. It
produced 'a new class of small villagers who cultivate the soil as
a subsidiary occupation, but find their chief occupation is industrial labour, the products of which they sell to the manufacturers
directly, or through the medium of merchants'.[5]

Primitive accumulation was thus associated with manufacture. Primitive accumulation was meant to accomplish the sep-

aration of the labourer from his means of production, but did not necessarily mean removing him from the countryside. For, as Marx argued, the merchant capital which commissioned a number of immediate producers provided the 'soil from which modern capitalism has grown', and here and there it still forms an aspect of what Marx called the 'formal subsumption of labour', that is, where capital took over an available and established labour process. The changes taking place in domestic industry during primary accumulation were a good example.

> The variations which can occur in the relations of supremacy and subordination without affecting the mode of production can be seen best where rural and domestic secondary industries, undertaken to satisfy the needs of individual families are transformed into autonomous burdens of capitalist industry.[6]

Elsewhere in *Capital* Marx also set out the definition of a stage in the development of the capitalist labour process which he termed 'manufacture'. Manufacture described a phase of handicraft workshop industry, a phase which preceded that of modern machine production. He was concerned with the organization and technological developments which set 'manufacture' off from previous industrial production. As a new form of organization, 'manufacture' described a workshop of handicraftsmen under capitalist control, carrying out one or a variety of tasks. As a new technology, it introduced the division of labour, though operations done by hand were still dependent on the skill of individuals and retained the character of a handicraft. Now what Marx actually intended to include in manufacture is a matter of some debate. He characterized manufacture as taking two distinct forms: heterogeneous manufacture, or the mechanical assembly of independently made components of the final product, as in the watch manufacture; and organic manufacture, or a series of connected processes, as in the manufacture of needles.

Marx made use of Babbage's proposition that manufacture would ideally display a fixed mathematical ratio between the size of each group of workers and each special function, and his model appeared to be a workshop in the charge of one capitalist.[7] Yet in the *Grundrisse* he also referred both to domestic industry and to centralized manufacture. Although Marx included rural industry under the manufacturing phase of capitalist production, he gave little consideration to the changes this rural manufacture might have entailed within the production process. He described how capitalist relations entered into rural production, arguing that manufacture seized hold initially, not of the urban trades, but of the rural secondary occupations where mass quantities were produced for export. But he also classified these occupations as examples of the social, not the technical, division of labour.[8] There is some debate over whether Marx meant to include centralized manufacture in his model, and it has recently been pointed out that as a model 'manufactures' has been related only to light industries. The heavy industries such as iron processing do not fit Marx's criteria of handicraft, manual labour and absence of machinery. Yet, equally, Marx does refer to ironworks, glass factories and paper mills as examples of industries which did fit his other criteria of mass production, marketing, investment and working capital on a large scale.[9]

It becomes clear that Marx was more concerned to point out the limitations of the system of manufactures than to detail the variety of its manifestations. He noted three basic inadequacies of the system. First, because a hierarchical structure was inserted into the division of labour the number of unskilled workers could not be infinitely extended. Hierarchy entailed the power and influence of skilled workers, and this prevented the full application of the Babbage principle.[10] Second, the narrow basis of handicraft itself excluded a really scientific division of the production process into its constituent parts. The division of labour could only go so far, for all parts had to be capable of being done by hand and forming a separate handicraft. The third and greatest problem, however, was the inability of capital to seize control of the whole disposable labour time of manufacturing workers.

Since handicraft skill is the foundation of manufacture, and since the mechanization of manufacture as a whole possesses no objective framework which would be independent of the workers themselves, capital is constantly compelled to wrestle with the insurbordination of the workers.[11]

The best historical example Marx could find to fit the criteria of his model of manufacture was the engineering workshop of the late eighteenth and early nineteenth centuries. He endorsed Andrew Ure's praise of the 'machine-factory' which displayed the division of labour in manifold gradations – the file, the drills, the lathe, having each its different workmen in the order of skill. There was also the special prophetic feature that 'this workshop, the product of the division of labour in manufacture, produced in its turn – machines'.[12]

In spite of the allusions to rural industry and centralized production, then, Marx's model of 'manufactures' seems to have been a large workshop in the hands of a capitalist and organized on the basis of wage labour. Though Marx clearly intended it to be an abstract model, he included many historical signposts. The image of a handicraft-workshop economy preceding the rise of the factory system has been readily accepted with little enquiry into 'manufactures' as a model rather than as an historical description. But it seemed so aptly to encapsulate the structures of some of the leading workshops and proto-factories of the day. There was the minute division of labour of the pin manufactory described in Diderot's *Encyclopédie*. The pin, the smallest and most common of all manufactured products, demanded perhaps the most operations before entering into commerce. That essay on the pin described eighteen different stages in the manufacturing process.[13]

There was also the division of handicraft skills in the Birmingham toy trades described in 1766:

There a button passes through fifty hands, and each hand passes perhaps a thousand in a day – likewise, by this means, the work becomes so simple that five times in six, children of 6 or 8 years old do it as well as men, and earn from 10 pence to 8 shillings a week.[14]

The Boulton and Watt works in 1790 seemed the epitomy of the order, regularity, and systematic layout emphasized by Marx.[15] Other manufacturers, such as Robert Peel in his Bury calico printing works in the 1780s, deployed unskilled labour under close discipline and supervision in much the manner described by Marx.[16]

These examples have frequently been accepted as indicative of the character of manufacturing organization in the country as a whole; or other manufacturing systems, such as 'putting-out', have been analysed purely in terms of 'manufactures', and both are regarded as the first of that two-step phase by which workers were deprived of their control of product and process. The minute division of labour in manufactures was then followed by the second step, the centralized organization of the factory system. 'Manufactures' was thus an innovation in organization, but one which obviously paled by comparison with the later innovation of the factory system.

'Manufactures' is credited with giving the capitalist, rather than the worker, control of the product, while the factory achieved this control over the production process itself.[17] This view of the system of manufactures is obviously 'retrospective'; its vantage point progressive modern industry. The linear framework of the model of manufactures has been incorporated into historians' discussion of industry before the factory system; it has also significantly affected our histories of workers' resistance. For just as the history of production was divided into control over product versus control over process, so eighteenth-century workers' struggles were divided off from those of the nineteenth century. Under manufactures, it has been assumed that each worker or group of workers still controlled, in some degree, the speed, intensity and rhythm of work, while later, under the factory system, machine-based modern industry was more profitable because it succeeded in taking away this control.

The model of manufactures was useful in highlighting the features of some of eighteenth-century industry, but it was a model and as such excluded the complication and variety of production processes. It was also a linear model, looking forward and back and not to either side, thus failing to place this manufacture in its own wider historical context. The clearest instance of both these problems is the model's failure to deal adequately with the features of the putting-out system and other related domestic forms of manufacture. This neglect of the putting-out system did not, however, prevent economists and historians from applying Marx's model. Maurice Dobb, in particular, presented a clear and provocative account of the domestic system.

> The domestic industry of this period, however, was in a crucial respect different from the gild handicraft from which it had descended: in the majority of cases it had become subordinated to the control of capital, and the producing craftsman had lost most of his economic independence of earlier times. . . .
>
> The craftsman's status was already beginning to approximate to that of a simple wage-earner; and in this respect the system was much closer to 'manufacture' than to the older urban handicrafts. . . . The subordination of production to capital, and the appearance of this class relationship between capitalist and the producer is, therefore, to be regarded as the crucial watershed between the old mode of production and the new, even if the technical changes that we associate with the industrial revolution were needed both to complete the transition and to afford scope for the full maturing of the capitalist mode of production. . . .[18]

Proto-industrialization: the theory and its problems

Where Marx himself did not dwell on the domestic system as the key manifestation of early modern manufacture, a recent historical school has identified rural putting-out systems with a distinct historical phase which preceded and paved the way for industrialization proper. Historians now look to the country cottage rather than the urban workshop for the crucial transitional phase of economic development, a phase now popularly known as 'proto-industrialization'.

Economic historians have long recognized the existence and importance of the great increase in commercial manufacturing production in the countryside between the seventeenth and nineteenth centuries. This rural industry practised in conjunction with agriculture has now, however, been elevated into the matrix of early modern economic and social change which paved the way for the factory system and wage labour, in short industrialization at a later date. Developments from the seventeenth to the nineteenth century are encapsulated in the following key changes. The world market for mass-produced goods grew at such a pace from the later sixteenth century that traditional urban manufacturers could not efficiently respond, hampered as they were by guild restrictions and high labour costs. Complementary agricultural development entailed increasing regional differentiation between arable and pastoral regions. A regional symbiosis based on comparative advantage ensued. An underemployed peasantry in pastoral regions became the basis for an expandable and self-exploiting industrial labour force, and the industry it took up improved the seasonal employment of labour. The possibilities of alternative industrial employment released the traditional limits placed on population growth by the size of landholdings. Rural workers, living as they did in a world of traditional peasant culture and values, took less than the customary urban wage for their industrial work and laboured more intensively in face of falling wages. In theory they had access to agricultural work which allowed them to produce part of their own subsistence. Their dispersal across the countryside furthermore made it difficult for them to organize so as to prevent price reductions by merchants. Access to this cheaper labour force therefore gave merchants a differential profit, one which was above the usual urban rates. This differential profit in turn provided a major source for capital accumulation. Proto-industry is credited not just with the sources of labour and capital, but in addition with the entrepreneurship and the technical and organizational changes that led to the first major increases in productivity before the factory.[19]

It is important to note that the emphasis given in this theory to cottage manufacture and putting-out systems is nothing new. Its proponents acknowledge their debt to later nineteenth-century German authorities such as Sombart, Troeltsch and Schmoller, and to the classic English economic histories by Unwin, Wadsworth and Mann, and Court. And recent English economic historians of the seventeenth and eighteenth centuries had already drawn attention to the demographic, agrarian and social organizational implications of the spread of rural industry.[20] But unlike this work, proto-industrialization implied systematic theory and predictive hypotheses.

If we wish to spell out more fully the definition of proto-industrialization, we can look to the criteria drawn up by Franklin Mendels who coined the term itself:

(1) the unit of reference is the region;
(2) rural industry in the region involved peasant participation in handicraft production for the market. Thus industry was seasonal and provided an income supplement, though it could ultimately be a full-time family occupation;
(3) the market for proto-industrial goods was international, not local;
(4) proto-industrial manufacture developed in symbiosis with commercial agriculture;
(5) towns in the region provided a locus for marketing, finishing and mercantile activity.

This definition of proto-industrialization went with a series of hypotheses.

(1) The higher incomes derived from handicraft production led to population growth, breaking the balance between labour supply and local subsistence; that is, handicraft generated the labour supply of the Industrial Revolution.
(2) Population growth and proto-industrialization soon generated diminishing returns, prompting changes in organization as well as new techniques which saved labour. In other words, proto-industrialization created pressures leading to the factory system and to new technology.

    (3) The profits from proto-industry accumulated in the
hands of merchants, commercial farmers and landlords;
that is, proto-industry led to the accumulation of capital.
    (4) Proto-industry required and generated specialist
knowledge of manufacturing organization and commerce; that is, proto-industry provided a training ground
for and a new supply of entrepreneurs.
    (5) Proto-industrialization and regional agricultural
specialization go hand in hand; that is, proto-industrialization leads to agricultural surpluses and reduces the price of food.

Proto-industrialization is thus credited with creating the key
changes in the use of land, labour, capital and entrepreneurship
which made the Industrial Revolution possible.[21]
The most widely explored aspect of Mendels's original theory
has been the demographic. Incentives to the emergence of
domestic industry, it is argued, were strengthened by population
pressure and magnified by the existence of partible inheritance.
Simultaneously, income-earning activities such as handicraft
production outside of agriculture released the traditional social
controls on marriage – inheritance and patriarchal control.
Mendels argued on the basis of evidence from Flanders that
earlier marriage and rapid rates of population increase were
associated with areas of proto-industry. But the demographic
evidence, despite the energy devoted to collecting it over the
past decade, has not given us the clue to causation. Did the
characteristics of proto-industry generate a growth in
population, or was higher population the factor which attracted
rural industry to an area?[22] Not only were the connections of the
model tenuous, but the results of its application to explain the
economic history of various regions have proved enormously
variable. Mendels called his proto-industry the first phase of the
industrialization process. Yet, in fact, for every area that made a
successful transition to the factory system – Lancashire,
Yorkshire, Lille, Alsace, the Rhineland, Saxony – there were
many more which headed straight for de-industrialization – the
West Country, East Anglia, the Cotswolds, rural Warwickshire,

Bedfordshire, Ulster, Brittany, Flanders, Silesia, Languedoc, Bavaria and Bohemia.

The discussion of these larger macroeconomic implications of proto-industry has complemented a debate at another level on the microeconomics of the proto-industrial manufacturing unit. The most sophisticated work from this new perspective has questioned the meaning of the division of labour in proto-industry. It has shown that there was little marked division of labour in the proto-industrial household. Several members of one household might carry on the same operation side by side, or alternatively individual processes might be put out to separate households. It was 'not the rule' for a proto-industrial household to be a 'miniature factory'. It was difficult to get a proper ratio of numbers employed in the different operations in the small unit of the proto-industrial household. The most important form of the division of labour was not the technical division of labour appealed to by Marx and the nineteenth-century economists, but a division of labour in the society and economy at large, that is, the specialization of regions in the mass production of a small number of articles.[23]

The sexual division of labour characteristic of such rural industry was also different from that imagined in more traditional literature. The characteristic household structure and production unit was not dominated by a servant class, but by the nuclear family which had recently turned to early marriage and high fertility to meet the needs of a new production process reliant on child and female labour. This household production unit reveals not a division of labour but the reverse – a new lack of separation between the work of men and women. As Hans Medick has put it, 'proto-industry brought the man back to the household'.[24]

The main reason for the success of this system of capitalist production, it is argued, was the built-in tendency to workers' self-exploitation. Even where guild artisans might mass-produce a commodity, they could not compete with rural labourers who had to be content with lower wages, both because they lacked corporate protection and because they had access to cheaper food. The higher profitability of rural industry did not necessarily imply higher productivity; if anything, productivity probably stagnated in the countryside until the connection between proto-industry and agriculture was weakened.

Proto-industrialization is used by historians as a phase or stage of development in much the same way as is 'manufactures'. Like 'manufactures' it, too, was said to 'contain the seeds of its own destruction'. For the same limitations of the high marginal costs of geographical dispersion and lack of regulation over work rhythms and quality pushed the putting-out system either into full factory production or into de-industrialization. Recent research has, however, diverged from Mendels's former unilinear framework. Historians have enquired into the reasons for the different outcomes of proto-industrialization. And they have tried to explain these by differences among regions in their internal economic, political and institutional environments, as well as in patterns of world trade. Although it is now accepted that the experience of proto-industrialization varied from region to region, the explanations given for this are based on two conflicting theories, one a theory of comparative advantage; the other a theory of noneconomic motivation.

Mendels and Jones explained the different regional specializations of Europe in terms of comparative advantage. According to Jones, comparative advantage accounted for the emergence of the North and Midland areas of England as the major manufacturing counties in the eighteenth century, while the southern regions de-industrialized and turned to the higher returns to be gained in agriculture. Areas of light and heavy soils adapted differently to the new crops and rotations of the Industrial Revolution, revealing a higher comparative advantage in agriculture in the light-soil regions of the South.[25] Similar regional differentiation is said to have taken place on the continent. But this mode of reasoning is really very unsatisfactory. Regional specialization and centres of proto-industry are explained by results – areas became proto-industrial because they did – and because they were good at what they did.

To this theory of comparative advantage is joined a simultaneous belief in the role of nonmarket behaviour or pre-industrial values. The emergence, success and ultimately the limitations of proto-industry are all explained by these values. And historians also appeal to these to explain away the continued existence of dispersed manufacture after the factory system. Proto-industry sometimes led into de-industrialization, it is argued, because of the subsistence orientation of workers.[26] And economic historians have long explained away that age-old *bête noir* of the success story of the Industrial Revolution – the handloom weaver and sweated industry – by appeals to the values of handicraft workers or family production units. The values of the family economy are thus left to account for a great deal, but we know little of what these values were.

Yet it is precisely these nonmarket values, those unanalysed areas of custom and culture, which determined the way in which individuals, families and communities reacted to new economic settings and constraints. This is not to say that they dwelt in a nonmonetary world or 'moral economy' which simply clashed with the market values of commerce and industry when these came to the countryside, but rather that they participated in this market with a different code of rationality, consuming at times when 'economic man' saved, playing at times when 'economic man' worked. A recognition and analysis of this plebeian culture is vital to the understanding of the dynamic of this period of 'manufactures' or 'proto-industry'.[27]

Manufacture and proto-industry: applications?

If we now look in more depth at the problems raised by the way in which the model of proto-industrialization has been used, we find them mirroring those of the model of manufactures. Both models assume the factory to be the ultimate method of organizing labour, and modern power-based machinery to be the best practice technology. The arrival of both made the Industrial Revolution, and they apparently eclipsed all other forms of technology and organization. But what do we know of these other technologies and manufacturing structures? Eighteenth-century manufacture was practised in all manner of different settings; it was organized along many different lines, each of which was 'rational' or legitimate in its own environment. Putting-out systems coexisted with artisan and cooperative forms of production, and all of these systems frequently interacted with some type of manufacture or protofactory. This was often the case within any one region. There was, for example, the Kentish Weald in the sixteenth and seventeenth centuries. It had a rural textile industry organized on a putting-out system which employed peasant labour, but it also had an important iron industry organized in centralized units around water-powered blast furnaces.

Even within one industry, these diverse forms of organization prevailed at the different stages of production. In eighteenth-century west Yorkshire, small artisan clothiers built and used their own cooperative mills for some of their preparatory processes. In eighteenth-century Lancashire, the Peels centralized their calico printing and spinning establishments, but ran extensive putting-out networks among weavers. In eighteenth-century Birmingham, small artisans in the hardware trades gathered together to build a centralized processing unit which supplied their brass and copper, and they 'put out' the production of parts and pieces in much the same way as did the lockmakers of the West Midlands and the nineteenth-century watchmakers of Coventry. How, then, did these many systems of manufacture succeed for so long in organizing work?

For an answer to this we might turn first to the economics of work organization. For it is here that we find some systematic attempt to compare the efficiencies and advantages of the different types of work organization at any one time. Williamson, who compared several different modes of organizing a batch industry such as pinmaking, assessed each mode in terms of its efficiency and its socioeconomic implications. Putting-out systems were compared to cooperative forms, and both in turn to capitalist subcontracting and factory hierarchies. On such assessments the factory system still appeared 'superior' in terms of efficiency to the putting-out system, and the putting-out system was better in turn than artisan ownership. The putting-out system had key disadvantages as compared to factories: high inventories, high transportation costs, poor work intensity, embezzlement and poor quality-control, and poor adaptation to sudden changes in markets or technique. In turn, the putting-out system has been deemed on such assessments to be preferable to artisan systems, for putting-out allowed the diffusion of knowledge of new materials or mixes of materials, and it ensured better quality-control of a standardized output. These advantages compounded cost advantages associated with the exchange of materials and the final product.[28] The economists of work organization, for all their claims to independent judgement of a wide range of forms of work organization, ultimately confirm the old linear framework of artisan stage succeeded by putting-out stage, with both finally eclipsed by the factory. Marx and the historians of proto-industry adopted the same framework.

Marx's analysis of the 'system of manufacture' and his historical view of production in the period just prior to the Industrial Revolution were mainly confined to the large workshop deploying the division of labour; while the proto-industry model has in general been applied only to the rural textile manufacture deploying putting-out systems. The large, hierarchically organized workshop was seen by Marx to be the most advanced form reached by manufacture before the limitations of the system became obvious. The historians of proto-industry have similarly regarded putting-out in such a light. They have certainly recognized that putting-out was not the only way of organizing industry before the factory. Indeed, they have distinguished between the 'Kauf system' (or artisan production) and the 'Verlag system' (or putting-out), but only to point to putting-out as the superior and dominant mode of organization before the factory system. It is back to Clapham's Industrial Revolution that we must go for any understanding of the polymorphic nature of industrial organization in this period. Clapham pointed out in 1930 that the Britain of a hundred years before had abounded in ancient and transitional types of industrial organization. While putting-out prevailed in much of the Scottish linen industry, in Dundee the spinners dealt directly with the manufacturer. While the woollen industry of the West

Country and the worsted industry of west Yorkshire were
model examples of putting-out, the woollen industry of the West
Riding was the seat of that independent artisan production
which received the praise of Defoe, Josiah Tucker and David
Hume. And the survival of these small independent clothiers
was ensured well into the nineteenth century, when in face of
the advantages of machinery and concentration in some processes they formed the cooperative or company units of which
Clapham and more recently Pat Hudson have written.[29] When
factories were forming in Lancashire, we must also remember
what Faucher said of Birmingham in the 1830s:

> The industry of this town, like French agriculture, has got
> into a state of parcellation. You meet ... hardly any big
> establishments: ... whilst capitals tend to concentrate in
> Great Britain, they divide more and more in Birmingham.[30]

Other alternatives: artisans, cooperatives and centralized manufacture

If we look at two alternatives to the putting-out system – artisan
production and centralized manufacture or protofactories – we
find many instances of their widespread use and success. In
other European countries and in earlier centuries artisan production had emerged as an alternative to both the medieval
guilds and the putting-out systems dominated by merchant
capital.[31] In an urban context artisan structures or small commodity production also developed their own dynamic,
sometimes alongside old guild-dominated production or as a
form of production appropriate to unincorporated towns and
areas. In sixteenth- century Leiden and Lille, 'small commodity
production' was not a 'stagnant hold-over from earlier days
found in declining sectors'. It was instead

> a system appropriate both to growing, market oriented in-
> dustries and to traditional urban societies. Competition and
> investment were intrinsic to this mode; at the same time, they

were circumscribed in accordance with certain firmly defended values. Full employment, a 'reasonable' standard of living, producer autonomy and rough equality among artisans rather than unbridled growth and profit maximization were the goals of the system.

The system incorporated constant pressure for innovation and change, but such innovation did not become the means for undermining this small commodity production. Urban manufacture did not necessarily shelter behind the protective walls of guild production or 'industrialize' to factory-based production. In areas like those of Leiden and Lille it developed its own artisanal system, a system which 'was not simply transitional or intermediate but formed one of the "obstacles" to the rise of capitalism'.[32]

Artisan-organized production was an equally dynamic industrial structure of the urban villages, suburbs and unincorporated towns of eighteenth-century Britain, in areas such as Birmingham and the London suburbs. It was a system of production which was not constrained by guild regulations but, nevertheless, just as in sixteenth-century Leiden and Lille, it did not operate purely according to the dictates of market forces; rather it was mediated through artisan customs and values. These customs and values, even in a guild-free environment, were expressed through journeymen's associations, as in the *compagnonnages* of eighteenth-century France, or simply through those principles of mutuality and cooperation contained in the 'custom of the trade'.[33]

Yet we know virtually nothing of the place of artisan structures or of cooperative alternatives in industrialization. These have been by and large disregarded by historians, written off as primitive structures of prehistory or as utopian failures. Economists and economic historians have nearly always taken the side of the winners and written for them. Industry, like labour, needs its sympathetic historian, one who would rescue all those forms of enterprise other than the factory from the dustbin of history, just as E. P. Thompson rescued 'the poor stockinger, the luddite cropper, the "obsolete" handloom weaver, the "utopian" artisan and even the deluded follower of Joanna Southcott, from the enormous condescension of posterity'. And we should remember Thompson's words: 'Our only criterion of judgement should not be whether or not a man's actions are justified in the light of subsequent evolution. After all, we are not at the end of social evolution ourselves.'[34] Integral to many such artisan systems of production were cooperative systems which were sometimes resorted to in order to ensure sources of materials or to complete a necessary stage of production involving centralized or mechanized processes, as we found in the metal and textile industries. A tradition of cooperation was also started in many trades as a temporary expedient for dealing with cyclical fluctuations.

In our attempt to uncover the history of some of the neglected artisanal and cooperative structures, we might take our inspiration not from Marx or the proto-industry model, but from eighteenth-century observers. For as we have seen already, Adam Smith regarded putting-out industries as a major source of labour's exploitation, but with artisanship he credited the most favourable of working conditions. With artisan structures, too, were to be found particular forms of artisan discipline and technology.

Time was a discipline which structured the artisan's life to an enormous extent. He or she worked within the limits of set delivery times of raw materials, availability of assistants who might have a different time economy, set dates for markets and fairs, and the time patterns of other social and income-earning activities. It is also striking that the theory of proto-industrialization has shunned technological change, assuming static technologies before the eighteenth century. But the significance, ubiquity and flexibility of handtools like the stamp, press, drawbench and lathe cemented artisan industry in Birmingham in the eighteenth and nineteenth centuries. As a witness to the Select Committee on Arts and Manufactures wrote in 1824,

> Our Birmingham machines are rarely, if ever mentioned in the scientific works of the day. The Birmingham machine is ephemeral . . . it has its existence only during the fashion of a

certain article, and it is contained within the precincts of a single manufactory of a town.[35]

And the inventions, improvements and adaptations made by these small artisans were often kept secret, incorporated in the special skill which guaranteed their craft superiority.

Apart from these artisan and cooperative manufacturing systems, which flourished alongside putting-out, there were those forms of industrial production which were centralized from the outset, as in mining and metal processing, and in the 'protofactories' which existed in the silk industry, in calico printing, in pinmaking, and in some of the factory colonies of the West Country woollen industry.[36] The calico-printing proto-factories have been described as the 'missing link' between the proto-industrial and modern industrial system in the textile industries. Such works were organized along the lines of Marx's workshop-manufactures, based on elementary labour-intensive techniques, the disciplining of the labour force, and the maximization of skills deriving from the division of handicraft labour.[37]

Although textiles formed the largest industry in the period before the factory, this does not mean that other industries organized around some central plant – as in mining, furnace, forge, brewing, distilling or boiling works – should be excluded from consideration as proto-industrialization. Many such activities entailed some change for better or worse in the local economy. The existence of centralized plant and processes did not prevent a seasonal or even family division of labour between industry and agriculture in much the same manner as in the textile putting-out industries. The same thing occurred frequently in mining and early iron forges and furnaces. One industry, such as metalworking, contained classic 'proto-factories' such as the Crawley ironworks, and at the same time the extensive division of labour found in a putting-out framework as in Peter Stubs's Warrington filemaking business. There was the Bristol wireworks producing pins in a protofactory side by side with dispersed and impoverished nailmakers exploited in a highly developed putting-out system.

The textile industry, too, easily assimilated partly mechanized mills into a countryside dominated by dispersed man- and womanpower.

Diversity and change

This diverse range of manufacturing structures coexisted within and across industries; in addition, these structures were rarely static. Either they adapted to changing market conditions with more or less success in their individual industries and regions, or industries changed their organization not in a linear but more frequently in a cyclical pattern.[38] There was no one pattern and no one criterion for the choice of a type of industrial organization. Profitability and labour costs were important determinants for the development of each structure, but, as we have seen, in a transitional capitalist society they were not the only ones – custom, community and patriarchal discipline played at least as significant a role in artisan, cooperative and protofactory alternatives to putting-out. This range of industrial structures also implied a range of different types of labour discipline and technological change. Yet we also know little of these.

Not only must we learn more of these alternative manufacturing structures, but we must more clearly assess the accomplishments and difficulties of the putting-out systems so frequently equated with proto-industrialization. Just how adaptable, for one, was the putting-out system to changes in demand?

Fabrics, of whatever fibres, in England or elsewhere, readily became identified in pre-industrial manufacture with particular regions... conservatism of method could almost be described as an endemic disease. Although the system knew little fixed capital, it had much human capital, which means minds as well as hands; to re-tool a factory or scrap a plant is often much easier than to bring about a fresh approach in management or to retrain a workforce with inherited ways of doing things.[39]

Social values

To conclude: the period just before industrialization was characterized by a multiplicity of different organizational structures of manufacture. The resilience of these structures was determined by their own special adaptability to the market, but it was also significantly affected by a range of nonmarket values and institutions. The directions of technological change and the choice of economic structures were, in other words, partly dependent on these social values of domestic workers and artisans. The strength of such values reverberated in the resistance to factories and to mechanization, ultimately determining the location of much factory-based industry. In moving beyond the market we can also start to account for all those nonwaged activities in which a largely underdeveloped workforce engaged in order to make up a daily subsistence. Credit and debt, perks and customary rights, and even embezzlement intersected with the extra economic relationships of the various forms of community – familial, civic, as well as trade-based – which made up the texture of the working man and woman's daily life.

We must seek to delve into those interstices of economic and customary relationship, interstices which have been neglected in economic history. Primitive accumulation and proto-industry have drawn our attention to the problem and to the years of transition before the Industrial Revolution, but the research still remains to be done to give these years an historical existence of their own.

CHAPTER 4

Agricultural Origins
of Industry

Adam Smith believed that agriculture was the foundation of economic development. Historians of proto-industrialization have looked to the countryside for the most significant manufacturing development of the eighteenth century. What precisely was the relationship between agriculture and industry over these years, and what made this relationship so special?

Agriculture's subordination

The experience of technical and institutional change in the agricultural sector requires its own separate study, one which this book cannot attempt to offer. What we shall do here is to examine the extent of interdependence between the sectors. Historians are keen to point out connections between agriculture and industry, but they have generally treated agriculture as the passive partner; the receptacle of raw materials and labour, there to be exploited by the industrial sector.

The timing of the Agricultural and Industrial Revolutions has confused the connection we seek. For most of the great agricultural innovators such as Turnip Townshend, Coke of Norfolk and Robert Bakewell coincided with, rather than preceded, the great industrial inventors – Arkwright, Watt, Darby and others. Agriculture thus took second place, its pace of innovation only complementary to and much slower than industry's. But this subordinate place was created partly out of the now outmoded economic models previously absorbed by economic historians, and partly out of a vulgarized Ricardian perspective which pervaded much nineteenth-century economic

debate. The latter conveyed the impression that agriculture was
the ultimate limitation on economic growth. It was a view aptly
expressed by J. S. Mill:

> The improvements which have been introduced into
> agriculture are so extremely limited, when compared with
> those of which some branches of manufactures have been
> found susceptible, and they are, besides, so very slow in
> making their way against those old habits and prejudices,
> which are perhaps more deeply rooted among farmers than
> among any other class of producers, that the progress of
> population seems in most instances to have kept pace with the
> improvement of population . . . It has not, hitherto, indeed
> been at any time the effect of an improvement to drive capital
> from the land, nor consequently to lower rent.[1]

This kind of anti-agrarian prejudice has produced an
enormously mistaken view based on a rather insignificant place
for agriculture in British industrialization. Indeed, as noted in
Chapter 1, the early decline of the relative importance of
agriculture in the distribution of national income and of the
labour force reflected as much an increase in its own
productivity as it did the very rapid progress of industry.

Recent research has now established that most of the key
technical and organizational changes in agriculture took place in
the seventeenth and early eighteenth centuries and not later.[2]
Agriculture, in fact, played a dynamic role in the wider
economy, for early and inexpensive technical improvements
increased labour productivity and made it possible to produce
much more food at lower unit costs. New population estimates
also demonstrate a sustained rise in population, not just from
the middle of the eighteenth century but from its very
beginning. And statistical evidence, which if anything
underestimates the improvement, shows that the most
substantial increases in output probably took place in the first
half of the eighteenth century, not the last half.[3] This chronology
of improvement now makes a strong case for the close
interdependence of agriculture and manufacturing, with the

springs of much manufacturing improvement to be found in the early dynamism of the agricultural sector.

Contributions to industry

Agriculture is generally considered to have made four types of contribution to industry. In the first place, it created a food surplus to feed a growing and increasingly urban population. Secondly, it helped to widen home and foreign markets. Thirdly, it provided a source of capital. And finally, it helped to condition a labour force and acted as a source and training ground for early management. The achievement of feeding Britain's rapidly expanding population for most of the eighteenth century never ranks among the success stories of the Industrial Revolution. This is because the technological breakthrough which allowed this was so nebulous and undramatic. Yet it is estimated that in the period 1650–1750 London's population rose by 70 per cent, expanding its proportion of national population from 7 to 11 per cent. In order to feed this population as well as that of the rest of the country, and still to export corn, output would have had to rise by at least 13 per cent in the period.[4] While population in England and Wales rose from 5.29 million in 1700 to 7.57 million in 1780, corn output in the same interval increased from 13,293,000 quarters to 16,106,000. Not only this, but corn was actually exported right up until 1780, the first year which saw substantial imports to the amount of 238,000 quarters. Population rose again to nearly 12 million by 1820 with corn output keeping step by rising to 25,086,000. By this year, however, corn imports had risen to 2,112,000.[5]

Agricultural surpluses

The sources of this increase in output lay in the agricultural innovations of the seventeenth and early eighteenth centuries – the irrigation, drainage, multiple crop rotations, and fodder and root crops described by others. These along with concomitant changes in agrarian structure – the decline of the peasantry and the rise of large-scale capitalist farming on a landlord-tenant basis – certainly seem to have raised the agricultural surplus. What effect did this then have on manufacturing industry? Its main effect was to release the old pre-industrial clamp of land limits and population pressure. Formerly there was little scope for the development of manufacturing, as diminishing marginal returns on the land placed an ever recurring barrier on both sustained population expansion and new patterns of demand. Technical progress in agriculture effectively extended the area of agricultural production, allowing real wages to remain the same or even rise in spite of population pressure. Higher real wages might indeed be spent on a greater number and range of consumer goods and not simply spread over more and bigger families. For productivity gains in agriculture enabled a smaller percentage of the labour force to feed the whole, and labour could be released into manufacturing, trade and distribution.

But in effect, it was not labour which was released, but labour time, for men and women moved only partly out of agriculture and soon found, in a combination of rural manufacturing and extensive cultivation, new economic incentives to increase their numbers. Innovations in agriculture, in addition, ultimately favoured grain-importing areas. For the same volume of manufactured goods, these areas could gain more foodstuffs and, therefore, support a higher population.

Markets and trade

Apart from releasing the British economy from the locks of a pre-industrial economic cycle, agricultural improvement provided a more precise foundation for industry through its impact on trade and on capital and labour markets. The food surpluses which allowed cities and industries to grow in turn created the mechanisms of trade and specialization which widened home and foreign markets. The precise connections and timing of the relations between agricultural improvement and greater demand for industrial commodities are difficult to specify, for consumption in the eighteenth century is a subject based on hypothesis and impressionistic evidence. Historians debate whether the home market was stagnating or buoyant in the first half of the eighteenth century. They are clear that agricultural prices fell steadily over these years. But whether this was due to a series of good harvests or to agricultural improvement, or simply to limited population pressure, is unclear. The sustained effect of agricultural improvement is now accepted as very significant. But the effect, in turn, of these years of cheap grain on the market for industrial commodities is still a puzzle, for the higher real incomes, which cheap grain implied, could be spent on more and better food, or on manufactured commodities. Recent estimates give more weight to the former than was previously thought. Good harvests, furthermore, affected the distribution as well as the level of incomes, and these could have moved in opposite directions. There seems to be no clear hypothetical relationship between food prices and industrial demand. Data linking food prices and industrial output in the early half of the century do indicate that falling food prices in years of peace were accompanied by rising industrial output, but in the last half of the century rising food prices and industrial output went hand in hand.[6] The effect of agricultural improvement and good harvests on rural incomes, and with this on industrial demand, needs a more critical approach. The data indicate some association, but the effect of a favourable conjuncture of harvests, agricultural change and population growth still seems open to uncertainty.

A first objection relates to the effect of this conjuncture on the distribution of income as a whole and is one that is fairly easily accommodated. The second questions the real gains of the labouring poor and is much more difficult to answer. First we must look at the differential effect of the harvest on the various landed classes and on the relations between town and country, agriculture and industry. As the great decline of the English peasantry had already happened by the 1730s and the greatest proportion of land was worked by tenant farmers paying money rents, we must also consider the effect of the harvest on these farmers and on their landlords, for their expenditure patterns were very different from those of the rural poor. The immediate impact of the harvest would be faced by tenant farmers who experienced declining terms of trade with industry. If wages were based on subsistence, lower food prices might mean lower labour costs, but not if wages were inflexible, for the result might then have been fewer working hours and therefore lower total income for farmers.[7] But notwithstanding this there must be counted an expanding middle-class market which maintained its momentum even against the erosion of the higher food prices which came in the latter half of the eighteenth century.[8] This market and the higher urban-industrial demand for foodstuffs which it simultaneously encouraged also prevented agricultural prices from becoming fatally depressed when higher agricultural productivity did encourage a downward trend of prices.[9]

An equally important connection between agriculture and industry was established through the landlords, for the movement of property income affected urban and rural industry alike. Cities and their industries might prosper in spite of high food prices and the lower discretionary incomes of the poor, for they were dependent on property incomes which rose with higher rents. But rural crises could also be transmitted eventually to the towns as landlords faced greater difficulty in collecting rents and struggled to maintain property incomes, while the state also faced problems in collecting taxes.

The problems of the countryside would be exacerbated by a constant syphoning off of the surplus to the benefit of the towns and by unproductive expenditure on various forms of conspicuous consumption. Attempts to maintain property incomes during such reversals would deprive both agriculture and rural industry of capital and incentives. The limitations inherent in these rural-urban connections were broken, however, by the introduction of convertible husbandry. For the new agriculture both created greater food surpluses and called for more investment in enclosure and transportation.[10] The local impact of these improvements was to drive away industry to other areas, yet it was also supposed to have raised the demand for wage goods from other areas, because the wage labourer as well as farmers, traders and town-dwellers bought their necessities and decencies instead of making their own.[11] But the extent to which this took place tells us not about rising rural living standards, but about the incursion of money into the village economy.

The conventional perspective on the relation between agricultural improvement and the expansion of trade has hinged on new rural expenditure patterns arising from that period of good harvests and agricultural improvements in the middle of the eighteenth century, characteristically dubbed 'The golden age of the labourer'. But before accepting this perspective we must know more of the conditions of the rural poor and something of the impact on *this* community of the composition of the market itself. By the eighteenth century it was already the case that most rural inhabitants were at least partly dependent on money wages. But new food surpluses did not necessarily bring better conditions or even higher wages, because the results of improvement varied according to the different agrarian economies. As Hoskins has shown for the village of Wigston in Leicestershire, the agricultural improvement of the period went hand in hand with agrarian reform, which brought money but not wealth to the community.

> The domestic economy of the whole village was radically altered. No longer could the peasant derive the necessaries of life from the materials, the soil, and the resources of his own countryside and his own strong arms. The self supporting peasant was transformed into a spender of money, for all the things he needed were now in the shops. Money which in the sixteenth century had played merely a marginal, though a necessary, part, now became the one thing necessary for the maintenance of life. Peasant thrift was replaced by commercial thrift. Every hour of work now had a money-value, unemployment became a disaster, for there was no piece of land the wage earner could turn to. His Elizabethan master had needed money intermittently, but *he* needs it nearly every day, certainly every week of the year.

The greater need for money at the time coincided, Hoskins shows, with rapidly rising poor rates.[12]

Historical judgement is also confirmed by contemporary observation. Eden in his extensive survey of the poor found that there was, if anything, a close connection between poverty and the market. Northern families preferred their subsistence economies to the shops, for though the clothing they produced was of a higher cost in terms of materials and time than most shop goods, it was also of higher quality. Workers in the Midlands and the South bought most of their clothing from the shopkeeper, and 'in the vicinity of the metropolis, working people seldom buy new clothes: they content themselves with a cast off coat, which may be usually purchased for about 5s. and second hand waistcoats, and breeches'.[13] And while markets did not mean wealth to the rural labourer, neither did they mean any connection between high real wages and local food surpluses. National and international corn markets, intercepted at various levels by myriads of corn middlemen, broke any local connection between the prices of food and the harvest. Not just the price of food, but the wages prevailing in most rural industry, were set by international not local markets.[14] Certainly the home market grew in the eighteenth century, but its expansion was based on changing social relations and not on a national trend of rising living standards.

This pessimistic interpretation of the growth of the market goes against the grain of much economic history of the period. But it tells the story of large regions of the country frequently neglected by economic historians. For Wigston was a Midland village, open-field, highly populated, and with a substantial proportion still holding land in 1750. It was slipping towards pastoral agriculture in the mid-eighteenth century, but until enclosure it contained a higher proportion of peasants than more arable areas of older enclosure and more consolidated landholdings. The latter areas had made their transition to 'agrarian capitalism' in the preceding two centuries.[15]

Just as the experience of the growth of the market was regional, not national, so too was the course of real wages which supposedly contributed to a widening market. The traditional division in the mid-eighteenth century between the earlier period of high real wages and the later period of falling real wages was tempered by the different agrarian and industrial economies between regions. Historians are now careful to point out, in addition, the frequent divergence between real wages and real incomes. Real wages were augmented, they argue, by higher labour inputs, through more intensive labour and more productive contributions from women and children.[16] But even the optimistic interpretation of eighteenth-century living standards remains equivocal. Wrigley and Schofield, on the basis of the Cambridge population data, argue that it is probably true that 'the secular tendency in standards of living as reflected in real wages was steadily upwards from the mid-seventeenth to the late eighteenth century, but that thereafter there was a sharp fall for about a generation before a resumption in the upward movement in the early nineteenth century'. The rising trend of real wages may have contributed to increasing nuptiality, but 'there are far too many uncertainties about the measurement of trends in real wages or in family purchasing power more generally to press the analysis very far'.[17]

Capital and labour
Equally tenuous is the conventional idea of agriculture's contribution to capital and labour markets. The narrow debate on the extent to which landlords invested in industry has really ended in stalemate. For while politics and prestige as much as profit influenced their business decisions, who can estimate the indirect effect?[18] Conventional dealings in the land market, in setting up trusts and obtaining mortgages, took on a new dimension in the commercial and industrial entries to the portfolios administered by provincial attorneys.[19] There seems, in addition, to have been no fixed division between agricultural and industrial capital. For many 'industrialists' moved their capital freely between the textile industry and traditional investment in the grocery trades, warehousing, agricultural processing and milling. The eighteenth-century Essex clothier Thomas Griggs alternated between the manufacture of textiles and the retailing of groceries; in addition he invested in real estate, fattened stock for winter markets, and malted barley for retail. Stubs, the Warrington filemaker, also worked the property markets, kept an inn and was a successful brewer.[20] Chapman argues that the records of 1,000 early eighteenth-century textile entrepreneurs from the West Country, East

Anglia and the Midlands show a close connection between agricultural and industrial investments. In these areas, unlike the West Riding which was known for its yeoman clothiers, this connection did not represent a direct involvement in both farming and manufacture. There was, rather, a direct connection between investment in textiles and investment in secondary agrarian activity. Many entrepreneurs divided their capital between the textile industry and the more traditional pursuits of malting, brewing or innkeeping in an attempt to limit their commitment to any particular industrial activity. The linkage existing between these agricultural and industrial pursuits formed a basis not just for the movement of capital from the land to manufacture, but also for the reverse movement. For when the older textile centres declined, assets were easily shifted back to agricultural processing, retailing and real estate.[21]

If the route to industrial capital formation from agricultural improvement was so convoluted, then the agricultural foundation for an industrial labour force was no less complicated. It is now widely argued that agricultural improvement was labour-using rather than labour-saving, so that most of the industrial labour force must have arisen out of population growth.[22] But historians are less easily persuaded to drop their clichés about the importance of the training and conditioning acquired by the English agricultural proletariat. First, it is not all that clear that local rather than national agricultural improvement and agrarian reform did increase employment. This view rests on a failure to take account of the newly pastoral regions. Furthermore, it concentrates on the labour-intensive nature of the enclosing process itself. This and the wartime movement towards arable production were both relatively short-lived phenomena. The employment levels of areas that converted to pasture plummeted. Wigston's labouring requirement for its 3,000 acres in 1832 was only forty labourers, about one quarter to a third of the labour needed in the old open fields.[23] But this unemployed labour force did not leave to flood new factories, nor did it, for that matter, become a fully fledged or even semi-proletariat.

The decline of Wigston's peasantry and its village went hand in hand with an increase in its population, for it became one of the many industrial villages which attracted the dispossessed from all round about to enter the domestic industry of framework knitting. The argument as put by E. L. Jones is that the workforce which engaged in by-employments 'acquired the technical expertise of industry'. Not only this, but it was being separated from the land and indeed from the ownership of all but a few tools and its own labour power. It worked for wages, bought its food and broke out of old customs of marriage and inheritance. It was a labour force which was being conditioned into that 'malleable and trained workforce' so 'central to an industrializing economy'.[24] But how accurate a picture is this of a very distinctive kind of labour force? Just as I have questioned the perspective on rural industry which sees it only as a problem of the transition to industrialization, so I would also argue that the existence of a rural industrial labour force does not provide any evidence of a direct role for agriculture in the formation of an industrial proletariat. The same historians who point to the modern characteristics of this rural industrial labour force also explain away its continued existence in the nineteenth century in face of the factory system by an appeal to its backward social values.[25] Agricultural improvement and agrarian reform did not create the makings of a 'semi-proletarianized' workforce on its way to the factory, but a rural domestic industrial workforce with its own internal dynamic. And where this rural industry did not emerge, unemployment was frequently the ultimate result of agricultural improvement. Even in those regions of arable farming, enclosure was frequently labour-saving, especially on larger farms such as the new estates created for tithe compensation.[26]

Agrarian structures and the rise of industry

Much the most significant connections between agriculture and industry were established via changing agrarian institutions and property holding. Different regional social structures engendered different industrial and technological experiences. Agrarian institutions hold the key to why industries arose in some areas rather than others, why they declined or moved to new frontiers, and why they adopted very different forms of organization. Joan Thirsk maintains that the rise of domestic industry is not to be explained by demography but by the existence of certain types of farming community and social organization.[27] The longer historical perspective shows that industrial by-employments were not particularly new for the seventeenth and eighteenth centuries. In mineral areas, for example, agriculture and industry were ancient by-employments – it is hard to say which started first. The reason for the rise of rural industry was not to be found in entrepreneurship, easy supplies of raw materials or even market demand, but in the economic circumstances of an area's inhabitants. The common factors in these circumstances were (1) a community of small freeholders or customary tenants with good tenure; (2) pastoral farming, that is, dairying or breeding; (3) no strong framework of cooperative agriculture, but rather equal division of land-holding.[28] Owners of circulating capital looked for labour in areas of weak manorialization which allowed immigration and the division of property among small cultivators. Areas of new settlement gave some livelihood to squatters, but areas of equal partition had also resulted in farms so small that the peasantry took to industry to supplement their small agricultural incomes. The Kentish Weald, northwest Wiltshire and central Suffolk provided examples of the first, and the Dales of the West Riding examples of the second, where.

> They knitted as they walked the village streets, they knitted in the dark because they were too poor to have a light; they knitted for dear life, because life was so cheap.[29]

The simple correspondence between areas of partible inheritance combined with pastoral agriculture and the location of the early domestic industries is, however, complicated by a number of factors. The seasonal interdependence of pastoral agriculture and domestic industry may have been a result rather than a cause of the location of rural industry. For the intensification of inter-regional economic competition in the

seventeenth and eighteenth centuries can explain industrial
location at least as well as the seasons can. Any divergence
between regions was accentuated by improvements on the
supply side of agriculture. Arable regions with surplus crops
'scooped' the urban markets, leaving less favoured regions to
turn to livestock and rural industry.[30] But if the precise direction of these agricultural influences is unclear, the differences
between the various systems of property holding and their implications for industry are very complicated indeed. The first
problem is that the division between partible inheritance and
primogeniture is a very blurred one. In spite of the minimal
difference in the end-result of inheritance by one system or the
other,[31] it does seem that partible inheritance tended to be
confined first to areas of sparse population, to forest areas with
the fringe benefits of forage and domestic industry, secondly to
areas of dense population supported by fishing or small industry,
or finally to rich pasture land. In areas like these the survival of
the family did not hinge on the distinction between land and
goods. But fielden arable areas shared little prospect of an
extension of land use, and there primogeniture supplemented by
cash legacies tended to dominate society. It was when the shift
was made from portions in kind to portions in cash that areas of
primogeniture became associated with large farms and large
cash portions; areas of partible inheritance went with dual
economies, poor farms and small portions.[32] The latter encouraged the expansion of rural industry; for most members of
the family, this provided the opportunity for making their small
stakes in the land a viable basis for subsistence. In areas of
primogeniture, however, younger sons who inherited cash
legacies and the skilled crafts of their fathers were freed from
the land to go and make their fortunes, and so became quite
different recruits for industry.[33] Though one of the common
features of many areas of domestic industry was weak manorialization or no strong framework of cooperative agriculture,
this is not to say that communal agriculture and industrial by-
employments did not coexist. Wigston was a case in point; so too
was the woollen area of the West Riding of Yorkshire.

For most cottagers and smallholders, however, the kind of

inheritance that was most important was not that of a tenure but that of use rights. David Hey has shown that it was not so much equal partition that was crucial as the existence of common rights, for these offered opportunities for squatters. Settlements near West Bromwich in the seventeenth and early eighteenth centuries consisted of small groups of cottages or ends around the heath, and the population kept cows and sheep and took up industrial employments. But the eighteenth-century influx of squatters and population growth brought encroachment on to the heath.[34] Indeed, common rights afforded the opportunity to take up by-employments. E. P. Thompson has described what he calls this 'grid of inheritance':

> The farmer, confronted with a dozen scattered strips in different hands, and with prescribed stints in the commons did not feel fiercely that he owned the land, that it was *his*. What he inherited was a place within the hierarchy of use rights; the right to tether his horse in the sykes, the right to unloose his stock for grazing, or for the cottager the right to get away with some timber foraging and casual grazing.[35]

The complications of the agrarian environment and the ultimate implications of these for industry make it difficult in the end to chart any simple general rules. There were certainly many proto-industrial communities to be found in environments which did not in any way fit the partible inheritance/pastoral agriculture model. The rural industries of the South of England – wool in East Anglia, pillow lace and straw plait in Buckinghamshire, Bedfordshire, Hertfordshire and Huntingdonshire, calico printing in Surrey, silk in Essex – were all set in areas of old enclosure and arable agriculture. The open-field arable villages of Leicestershire provided the setting for a knitting industry, and the arable Western Lowlands of Scotland held a substantial rural spinning industry.[36]

Different agricultural and landholding systems might be pre-conditions for purely agricultural or semi-industrial regions, but research has not yet established just what the connections were. They could as well help to explain the emergence of different

types of industry or at least of industrial organization. The West
Midlands metal industries of the early eighteenth century were
both highly specialized and localized. The scythesmiths were to
be found in the southern parts of the region close to the fielden
parishes of Worcestershire and Warwickshire. They ran
large-scale farms and large workshops containing as many as
eight anvils and six bellows. Their trade demanded more skill
and capital than the other metal crafts and did not lend itself to
putting-out arrangements, so that scythemakers retained all the
stages of production in their own hands. Lorimers and saddlers'
ironmongers were concentrated in Walsall where they held
small parcels of customary or freehold land in this relatively
urban community. The nailworkers had very small holdings and
practised their trade as a seasonal occupation alternating with
agriculture. Unlike the lorimers, who used their smallholdings
to raise mortgages in order to finance their industrial capital, the
nailers used their plots to eke out a subsistence which they failed
to maintain out of the measly piece rates paid by nail factors and
putting-out ironmongers.[37]

Different inheritance systems and the differential decline of
manorialism also help to explain the emergence of different
industries and different industrial organizations within the West
Riding of Yorkshire. The traditional woollen industry found its
context in an area of large fertile holdings dominated by
traditional manorial controls over land tenure. More copyhold
land was retained, and the commons were not enclosed until the
late eighteenth century. This industry was organized on a 'Kauf
system' basis; the independent farmer-weaver, not the factory
system or the putting-out merchant, remained the pillar of the
industry into the nineteenth century. The new worsted industry
by contrast emerged in an area of early enclosure, and en-
franchisement with partible inheritance, while economic and
social divisions between an increasing landless element and a
small elite of large-scale merchant capitalists, who operated a
putting-out system, paved the way for the factory system.[38]

Research would no doubt reveal more of the complexities of
relations between agrarian institutions of property holding and
industrial organization and technology. The land may hold the

key to the distinctive technological and organizational traditions of the different regions of Britain. The different coalmining technologies of Staffordshire and the Northeast, the method of wage payments peculiar to Cornish mining areas, the contrast between the industrial organization of Shropshire and neighbouring Staffordshire and other regional peculiarities noticed by Sidney Pollard, owe much to local traditions, and these traditions in turn find their context in the complicated local variations in the holding and transmission of landed property.[39]

These tantalizing connections between agrarian relations and industry can only be explored region by region. A link between changes in agriculture and the rise of manufacture exists, but we are no longer sure of the effects of agricultural change on the supplies of labour, capital and organization to new and older industries. As with agrarian relations, the strength of the link and its direct and indirect effects vary enormously between regions. This is not to dismiss the agricultural foundation of the age of manufacture, but to point to the complications of this foundation and the many different edifices that might be erected on it.

CHAPTER 5

Industrial Decline

Industrial transformation in the eighteenth and early nineteenth centuries meant not only the rise of new industries and the reorganization of the old. It also entailed the decline of old industries and erasure of old methods of production. Earlier chapters have pointed out the significance historians attach to the spread of the rural domestic industries in the sixteenth to eighteenth centuries, and the bridge these industries formed between agricultural and industrial development. Now we shall examine some of the failures of this rural manufacture or proto-industrialization. We shall trace the road to decline in some of these industries and production methods over the eighteenth century.

Joan Thirsk has described the growth and wide geographical dispersal of industrial by-employments in the seventeenth century. Most of these industries grew up within a pastoral economy, though some found a more suitable context in the towns. The manufacture of starch, needles, pins, cooking pots, kettles, frying pans, lace, soap, vinegar and stockings, as well as the more conventional iron, glass, brass, lead and coal industries, now employed large numbers of part- and full-time workers. Divisions in the quality of commodities coincided in many cases with divisions between town and country industries. The best knives were made in Sheffield itself; the inferior and cheaper ones in the villages round about. Coarse pottery was made in the country while delftware and creamware were crafts of the large towns. Woollen and worsted stockings were made in the country; jersey stockings in Norwich and London; and silk stockings only in London.[1] Some areas of the country were already industrialized regions by the seventeenth century. Staffordshire, for example, had only a small central zone with no industries practised alongside farming. There were wood turning, carpentry and tanning in the Needlewood Forest, coal in south Staffordshire, as well as iron and metal goods including locks, handles, buttons, saddlery and nails, coal and iron in Cannock Chase. Kinver Forest in the southwest had scythesmiths and makers of edge tools, and there were glassworkers on the Staffordshire-Worcestershire border at Stourbridge. Bursham in the northwest had a pottery industry, and there was ironstone mining in the northeast. Leather working and textile weaving in hemp, flax and wool were scattered throughout the country.[2] In Essex by 1629, 40,000 to 50,000 people were said to be wholly dependent on the manufacture of the New Draperies. These rural workers were 'not able to subsist unless they be continually set on work, and weekly paid'. A trade crisis in 1629 meant the 'instant impoverishment of multitudes of those poor people'.[3]

We know of the emergence of these industries in the sixteenth and seventeenth centuries, and subsequently of the factory-based textile industries and large-scale metal manufactures which formed the basis for industrialization in the late eighteenth and early nineteenth centuries. But we know little of the course of economic change between these two periods nor of what happened to all those industries set up by hopeful projectors before and after the Civil War. Here we will look at the forgotten side-effects of the greater regional concentration of industry and the development of factory production. Industrial decline is a subject curiously neglected by economic historians who are generally more concerned to point out the triumphs of industrialization and the growth of new regions. They have assumed that most regions of domestic industry went on to join the story of the factory system, and for those that did not do so immediately it was apparent that the effect of economic growth from the late eighteenth century onwards was to generate employment opportunities in hand-domestic as well as mechanized sectors. We are aware of a whole series of traditional domestic industries which were eclipsed in the course of the eighteenth and early nineteenth centuries, but know little of the course of their decline and nothing of the short- if not long-term effect of this de-industrialization. Sidney Pollard has admitted industrial decline as part of a European pattern of ebb and flow between regions,[4] and Eric Richards has postulated an association between the decline of a whole series of traditional domestic industries and a contraction of employment opportunities for women over the course of the nineteenth century. Richards argues that not only did employment for women in agriculture fall in the nineteenth century, but that women's hand trades withered away. The new cotton industry did generate jobs for women but these were concentrated in only a few regions, and other new industries created few employment opportunities. 'For women the gradual loss in employment in the traditional lines was probably greater than the creation of new opportunities.'[5]

The greatest decline of these traditional industries probably took place in the years just following the Napoleonic Wars. But a substantial revolution, if not actual decline, is also evident over the course of the eighteenth century. Historians who have acknowledged this decline ascribe it to a number of factors: a regional phenomenon reflecting changes in comparative advantage;[6] the failure of traditional entrepreneurship and labour;[7] or a specific outcome of cyclical downturn in the middle of the century.[8] Be that as it may, the most striking change in the industrial structure of the country between the end of the seventeenth century and the middle of the nineteenth was its geography. The urbanized industrial zone of the seventeenth century extended along a right angle connecting Bristol, London and Norwich. By the nineteenth century this zone had shifted north and northwest to the coalfields of the West Midlands, the West Riding, Lancashire and South Wales. The old manufacturing centres of southern and eastern England had languished or disappeared.[9]

While peasant society was mean and poor, the coming of some domestic industries could make the difference between destitution and decency for the poor and dispossessed. Defoe in the early eighteenth century reported on the living conditions of some of the leadminers of Bassington Moor in Derbyshire, where he found a woman and her family living in a cave.

The habitation was poor, 'tis true, but things within did not look so like misery as I expected. Everything was clean and neat, though mean and ordinary. There were shelves with earthenware and some pewter and brass. There was . . . a whole flitch or side of bacon hanging up in the chimney and by it a good piece of another. There was a sow and pigs running about at the door, and a little lean cow feeding upon a green place just before the door . . . a little enclosed piece of ground was growing with good barley.

The woman's husband worked in the leadmines for 5d a day, and when she could she worked ore for 3d a day. Defoe was amazed that 8d a day was enough to maintain a man, his wife and five small children, but declared that they 'seemed to live pleasantly'. The children looked 'plump, ruddy and wholesome and the woman was full, well shaped and clean'. He found nothing there which looked like the 'dirt and nastiness of the miserable cottages of the poor'.[10]

But other contemporaries saw different conditions, for such industry was as uncertain as the volatile trade which encouraged it. An observer in 1677 wrote that

though it sets the poor on work where it finds them, yet it draws them still more to the place; and their masters allow wages so mean that they are only preserved from starving while they work; when age, sickness and death comes, themselves, their wives or their children are commonly left on the parish.[11]

The variable conditions of these areas of industrial by-employment even in the seventeenth century were dramatically accentuated by the nineteenth. Some areas had declined into terrible poverty. Hoskins found that the peasant society of Wigston Magna, practically all framework knitters by the 1830s, lived in overcrowded streets, only to die at alarming rates of puerperal fever, consumption, and infant mortality. 'Wages were low, housing conditions worse than they had been since the sixteenth century, unemployment had become endemic.'

And the trade declined amidst all those depressing institutions of decline – truck payments, debts and proliferating middlemen. By 1845 knitters could rarely earn more than 7s a week, and young people were finally giving up this low-paid trade for a new one, sewing and stitching gloves.[12]

The decline of industries in the South of England took place earlier, but the effect on local communities at least in the short term was no less devastating. E. L. Jones has charted the grim litany of regional decline in Berkshire, Dorset, Hampshire, Wiltshire, Norfolk, Suffolk and Essex, areas which dropped an average of eleven places in the county league table of wealth between 1693 and 1843. Ironworking left Kent, the Sussex Weald, then the Forest of Dean in the course of the eighteenth century. The woollen cloth industry disappeared from Kent in the late seventeenth century and was finally eclipsed in the later eighteenth in Surrey, Berkshire and Hampshire. It dwindled in Exeter, and by the early nineteenth century was contracting in Somerset, Wiltshire and Gloucestershire. Neither carpet weaving, cotton spinning, nor stocking knitting sustained a hold in the south. By the nineteenth century boot- and shoemaking had disappeared from Berkshire, the making of wire buttons from Dorset, and wool and fur hatting from Gloucestershire.[13] Besides these regions, where industry contracted over the eighteenth century, may be placed others which saw new industries and innovations spurt briefly and retreat thereafter. Pollard names ten regions which 'saw significant innovatory change in the period between the 1760s and 1790s'. But of these he found that Cornwall, Shropshire, North Wales and the Derbyshire uplands fell by the wayside after making their vital contributions, and Tyneside and Clydeside had to find a 'second wind to survive as industrial centres'.[14]

Reasons for decline

What were the causes of these different regional experiences of expansion and decline? Many historians are quick to point to the inexorable emergence of 'comparative advantage' between industry and agriculture in various regions. Some regions were better suited to take up the new agricultural techniques of the period, while others had distinctive industrial advantages. Regions then specialized accordingly. Sidney Pollard charges a more specific set of factors with the subsequent decline of industrial regions including exhaustion of minerals, discovery of cheaper alternative supplies, locational shifts elsewhere, new developments in transportation which made the region less favourable than its rivals, or the fetter of small size.[15] We need to consider the course and causes of industrial decline in the case of several regions and industries, looking first at the most celebrated case of decline – that of the traditional cloth industries of East Anglia and the West of England. Some contemporaries attributed the failure of these regions to a lack of basic raw materials, for example coal and water power. Yet, as Jones has shown, the persistence of the blanket manufacture at Whitney was specifically attributed in 1809 to cheap local labour offsetting the 'want of vicinity to coal'. The Wiltshire mills had access to Somerset coal, and the competition of agriculture for water power in the South hardly accounts for the extent of market retreat.[16] As we shall see, more weight should be placed on the social and institutional factors, and we need to re-examine the meaning of comparative advantage as an explanation for regional growth and decline.

The old cloth regions

The cloth industry of Essex was the first major southern industry to collapse. In the seventeenth and eighteenth centuries it dominated the life of four large towns as well as a further dozen towns and large villages: it contributed something to the income of most Essex families. Defoe had seen

> the villages stand thick, the market towns not only more in number but larger and fuller of inhabitants and, in short, the whole county full of little endships or hamlets and scattered houses, that it looks all like a planted colony, everywhere full of people and the people everywhere full of business.[17]

By 1800 the cloth industry was gone. The two regions had specialized in says and bays, that is, cloth which was half worsted and half fulled. Rural weaving, which had been wide-spread, was the first to decline, disappearing soon after 1700. At this time the few surviving village firms all ceased, and weaving continued for a time in the towns. Spinning, however, was widespread through the countryside, and in the 1740s it was thought the majority of women were employed in it. In addition to the cloth industry fustians and cottons were manufactured across northwest Essex into Suffolk. Their coming had caused smallholders to settle in the area, growing woad for dyeing, and this survived until mechanization in the North at the end of the century. The cloth trade, however, collapsed even before mechanization posed any threat.

The first casualties were the small centres where most of the clothiers were master weavers with a small capital. The major centre was Halstead where long-established family businesses also maintained extensive interests in farming, malting and milling, following the classic pattern of pre-industrial capital formation. Most of the men were weavers and the women spinners. But apprenticeship suddenly declined after 1780, population fell soon after, and by 1791 it was reported that the industry was in a declining state: there were only four clothiers left, and by 1800 even these were gone. Other Essex cloth towns declined even earlier. The centuries-old cloth trade of Coggleshall, buttressed by very substantial business, was in decay by 1720. In 1733 it was petitioning Parliament for help; in 1740 there was a major poor law crisis; a short recovery was followed by collapse again in the 1760s. The early 1790s saw the last workers' procession and the dissolution of the Cloth Work-ers' Company. Even Colchester, known for its high standards and pre-eminent craftsmen, failed to escape the general malaise. A general decline from 1700, punctuated by occasional booms following on the end of wars, escalated after the 1760s to a general state of bankruptcy.[18]

The booms were memorable. Defoe remembered that 'after the late plague in France and the Peace in Spain the Run for goods was so great in England and the Price of everything rose so high that the poor women in Essex could earn one shilling to one shilling and sixpence per diem by spinning . . . the poor farmers could get no dairy maids . . . they all run away to Bocking, Sudbury, to Braintree and to other manufacturing towns in Essex and Suffolk. The very Plowman did the same . . .' But this prosperity was not long lasting, for 'As soon as the demand slack'd from abroad all these loose people were turn'd off, the spinsters went to begging, the weavers rose in rebellion.'[19]

Norfolk's industry was in trouble from the first decade of the eighteenth century thanks to the transfer of the stocking manufacture. There was a petition to Parliament in 1709 on the decline of trade, followed by riots against printed calico ten years later. Norfolk staggered under the blow of competition first from East India goods, then from Yorkshire worsteds, and finally from cotton. Prosperity returned in the 1750s and 1760s followed by decline again in 1765. Trade picked up in the 1770s, but was low again in the 1780s.[20] Stumbling between periods of stagnation and activity, the worsted industry did continue to grow until the 1770s, and though it saw the occasional light of prosperity until the 1820s, it was by then well on the wane behind the rising star of the West Riding.[21] By 1820 *Rees's Cyclopedia* could report that the manufacture of worsteds had transferred to Yorkshire and that the manufacture of camlets, calimancoes and bombazines had disappeared from Norwich.[22]

The cloth industry of the West Country held out until the early nineteenth century, but it was subject to fluctuating fortunes for most of the eighteenth. While output of broadcloth in Yorkshire doubled between 1727 and 1765, that of the West Country did not grow at all – it stagnated right up until 1770. The little improvement which followed was eclipsed by a severe depression in Gloucestershire in 1783–4, and the last five years of the decade in no way overcame this. A growth in demand in the early 1790s, however, encouraged manufacturers and workers to acquiesce in the use of spinning machinery. The industry grew in Wiltshire and Somerset, making great advances in Frome. But by 1800 there was more unemployment in Wiltshire and Somerset. An air of confidence appeared once more in Gloucestershire in the early 1820s. But except inTrowbridge the small clothier was replaced by the large factory, though outside handloom weaving continued in the West. The crisis of 1826 was the real dividing line, and while the industry picked up again in the early 1830s, it had actually shifted in location from the Upper Stroudwater Valley and the region below the Cotswolds to concentrate around Stroud and the Nailsworth Valley. But the West did not take up steam power; by the 1830s it had lost almost all its manufacture of cheaper cloth, and like Essex and Norfolk was experiencing population decline.[23]

The pattern and timing of decline in Essex, Norfolk and the West Country was also that of Suffolk, Coventry, Worcester, Dorset and Exeter. The serge industry of Exeter, which claimed an export trade of £500,000 in 1700, had ceased to be of any importance to the city by 1800.[24]

Comparative advantage

The explanations given for the decline of these old cloth regions are as varied as those for the rise of the industrial North. They range from geography to institutions, including the characteristics of management and labour. Though Clapham dismissed the significance of water power and coal as factors behind the shift of the industry to the North, a new form of natural determination has arisen based on the theory of comparative advantage. It is E. L. Jones's view that regions progressively discovered their comparative advantage in either agriculture or industry and increasingly concentrated resources in the one or the other. The kinds of agricultural innovation which occurred in the seventeenth and eighteenth centuries favoured the light land soils of the South, where agriculture became the more profitable activity. In the North and the Midlands, on the other hand, domestic industries gathered in areas suited only to pastoral husbandry, and these industrial occupations became relatively more profitable even before other advantages of coal and water power came into play in the later period of mechanization.

The North and some Midland districts became more industrial precisely because the readier uptake of the 'new' crops on the light land in the south had made them relatively poorer agriculturally. North and south thus evolved as complementary markets which it became worth linking by better communications.[25]

The decline of the Essex cloth industry was a pre-eminent example of this. Though the decline of the industry was blamed directly on its product specialization, there had to be some reason why clothiers did not introduce new product lines. Specialization in bays and says with limited markets in Spain, Portugal and Latin America made the link with Lisbon crucial, and the series of eighteenth-century wars was disastrous for trade. But in Essex the traditional dual occupations of farming and textiles formed the basis for some clothiers to strengthen their ties with farming, especially after 1700 when both the profitability and the social standing of agriculture improved.[26] It had also long been customary for West Country clothiers to buy estates and become gentlemen clothiers. But it was in addition evident that the spread of investment across industrial- and agrarian-related pursuits was part and parcel of the pre-industrial pattern of capital formation.[27] That resources were shifted over to agriculture and away from industry was nothing new. What was new was the extent and the timing of this shift of resources. Jones points out that in Gloucestershire this transfer of capital into land was said to have caused the failure of clothiers whose assets were no longer liquid enough to tide them over trade depressions. But more important was a prolonged and unfortunately timed withdrawal from industry into agriculture which seemed to coincide with the recognition that local industrial investment was no longer rewarding.

This argument seems to fit the experiences of Essex, Berkshire and Norfolk, which were in decline even before the threat of mechanization. Subsequently this was reinforced by the dictates of sources of power.

On the clays and lowland heaths of the south and east, with little or no alternative to 'mother and daughter power', no escape from the spinning wheel or hand loom, cottage industry contracted in the face of competition from machines. . . . Handicraft workers cut their prices to the bone in order to match machine production . . . the operators of water driven textile mills cut their prices when in turn they were also faced with competition from steam. Coal beat them. Without it areas with waterpower alone followed handicraft districts into industrial oblivion.[28]

Comparative advantage in agriculture cannot, however, provide a sufficient explanation for industrial decline in the old clothing districts. For it explains much of what occurred by the end result, failing to provide enough in the way of independent factors contributing to industrial decline. Reasons behind the original aptitude of regions for the one pursuit or the other are not explored. It is also clear that though the cloth industry in the southern regions did not entirely de-industrialize, it shifted resources into more profitable agricultural pursuits. It is clear that the capital and labour of the failing cloth industry also provided a congenial environment to attract a whole series of smaller domestic industries in silk, lace, and strawplait manufacture, glovemaking and shirt-buttonmaking, until these industries too faced the threat of the machine.

Institutional factors

If comparative advantage in agriculture is not reason itself for the shift away from the southern cloth manufacture, what other explanations are there? There were several notable institutional rigidities. The first was restrictions on capital and entrepreneurship. Control by the Blackwell Hall factors is said to have restricted opportunities for the small clothier. By manipulating credit the factors apparently split the clothiers into a small wealthy group and a large group of men with inadequate credit. Another factor was the existence of high and inflexible wage rates in the South. It is often argued that in the late seventeenth century the many competing occupations in Kent and the high price of food in the Home Counties so forced up wages that Kentish clothiers could no longer compete with lower wage areas.

But in the remainder of the southern cloth industry wages were by the second half of the eighteenth century lower than in the North. The final rigidity was the polarity of master and man in the South compared to the socially more uniform small weaver communities of the North. This produced more forceful workers' resistance to mechanization in the South.[29]

The limitations of entrepreneurship do seem to have played a contributory, if not decisive part in the route to decline. It was said of the West Country that it was founded on a monopoly erected and supported by great capitals and this led to conservatism. Most have accepted Josiah Tucker's well-known comparison of Yorkshire and the West Country. In Yorkshire

> Their journeymen . . . if they have any, being so little removed from the Degree and Condition of their masters, and likely to set up for themselves by the Industry and Frugality of a few years . . . thus it is, that the working people are generally Moral, Sober, and Industrious; that the goods are well made, and exceedingly cheap.

In the West Country

> The Motives to Industry, Frugality and Sobriety are all subverted to this one consideration viz. that they shall always be chained to the same Oar (the Clothier), and never be but Journeymen. . . . Is it little wonder that the trade in Yorkshire should flourish, or the trade in Somersetshire, Wiltshire, and Gloucestershire be found declining every Day?[30]

More recently several historians have made a similar observation on the deficiencies of entrepreneurship: 'All through the eighteenth century the Company [of Weavers, Fullers and Shearmen] was inbreeding, with the same families and their ideas preserved like heirlooms.'[31] Although, as Julia Mann points out, clothiers, small as well as large, abounded in the West Country in the eighteenth century, the conditions for the survival of the small man became more difficult over the period as divisions between 'respectable' and 'inferior' clothiers sharpened. Mann, however, blames the clothiers for being too easy-going, 'faithful for too long' to the older types of cloth. [32] Yet another failure lay in marketing; this explains the continued specialization in the same product lines and the same markets. For all the West Country cloth exports went through Blackwell Hall. There were no public cloth halls in the West Country or Norwich where clothiers could exhibit and sell their wares.

> Export outlets in the West Country were entirely, if somewhat indirectly through the Blackwell Hall factors, controlled by merchants in London. . . . They were not experts in handling cloth. The merchants in Leeds and Wakefield were a different species. Cloth was their life, their sole interest. . . . The difference between the ways in which the West Riding trade was handled by the active merchants of Leeds, Wakefield (and eventually Halifax) and the exports of every other production area from Norwich down, which were all monopolized by non-specialist London traders often working within the restrictions of the trading companies themselves, accounts in good measure for Yorkshire's growing supremacy in the eighteenth century.[33]

The social divisions which inhibited enterprise also provoked extensive workers' antagonism to technical and organizational change. Though decline in the South had set in before the time of water- and steam-powered factories, this is not to deny the role of earlier technical innovation. The examples of resistance to machinery are legion throughout the eighteenth century. In Essex class conflict was often acute. Weavers in the towns shared identical working conditions and were well placed to combine. The weavers' revolt of 1715 extinguished any idea of a factory system. There was a big strike in 1757 when Colchester employers demanded the return of thrums (the ends left on the loom after weaving). Weavers in Barking fought against the wool mill (for cleaning and loosening wool) in 1759. There was also some resistance to the flying shuttle, though it was introduced from the 1750s. Spinning remained backward, and there were no jennies until 1794. But cheap domestic spinning labour seemed a viable alternative, and it just did not seem profitable to invest in machinery at a time of imminent industrial collapse.[34]

In the West Country, weavers faced an almost continuous decline in wage rates at least until the middle of the eighteenth century. Spinners' earnings were more volatile, but when they fell a widely dispersed body of female spinners had no organization to marshal resistance. The wretched state of these spinners and other clothworkers in Gloucestershire was more often expressed in embezzlement and shoddy work. By the 1780s the clothiers of Minchinhampton sent all their wool to be spun outside the area, for 'Our poor spoil their yarn by dirtiness, bad spinning, dumping and frequently putting several workers' yarn together and many other frauds.'[35]

Spinning occupied women over a wide area in what were otherwise purely agricultural districts. Though there were spinning jennies and carding engines in Yorkshire from the 1770s, few were introduced into the West Country until the 1790s. A jenny set up in Shepton Mallet in 1776 was destroyed by a mob. Elsewhere until the 1790s it was only on the fringe of the industrial area that machinery was set up. The fierce resistance to the jenny at Keynsham among colliers and their wives, dependent upon the supplementary income from spinning, may have eroded the industry there, for, whatever the reasons, the industry left the area soon after.[36]

Within the West Country, Wiltshire and Somerset continued to be much more hostile to machinery than Gloucestershire. A mob destroyed an advanced scribbling machine in Bradford-on-Avon in 1791. Workpeople rioted against the flying shuttle in Trowbridge in 1785–7 and 1810–13, postponing its introduction there and in west Wiltshire until the end of the Napoleonic Wars. Weavers were still rioting against the flying shuttle in Frome in 1822. Resistance to finishing machinery was even more celebrated, in Yorkshire fuelling the classic Luddite attacks, and just as strongly resisted in Wiltshire and Somerset. Unlike the clothworkers of Gloucestershire who had long ago adopted the gig mill, those of Wiltshire and Somerset believed the machine would be the first step to the introduction of the shearing frame, so that it remained rare in these counties until the end of the wars.[37]

The greater resistance to the jenny in the South and the East can be accounted for by the much greater importance of hand spinning to the subsistence of poor women and rural families generally. There were fewer poor women dependent on spinning alone in the North, and anyway those previously employed in the wool trade could get employment well into the nineteenth century in spinning worsted, for which the jenny was little used. In the South and the East, it was not just machinery but the decline of the worsted trade itself which removed the mainstay for thousands of women and children in the villages of Norfolk and Suffolk. The concentration of industry did bring more employment and higher wages for women in the districts where mechanization was introduced, but it spelt doom to thousands of women in scattered outlying parishes.[38]

New rural industries for old?

But where the cloth industry declined in the South of England, other smaller domestic industries grew up, some of them continuing into the later nineteenth century. Comparative advantage in agriculture did not prevent the emergence of new industries there when the old cloth centres declined. However, it would be a big mistake to regard these industries as a replacement for the old cloth industries. In the first place, they were smaller and poorer than their great predecessor. And, in addition, they were sometimes not real replacements at all. In Northamptonshire, for example, shoes apparently replaced worsted, but in fact they were made in different areas and employed different sections of the country population.[39]

The decline of several Essex towns was stayed by the silk manufacture and by the substitution of bunting for the more expensive worsted previously made. This coarse cheap fabric continued to employ women in and around Sudbury until the 1870s.[40] The ancient bone and pillow lace industry in the early eighteenth century was the staple of women and children in Bedford, Buckinghamshire, Northampton, Devon, Dorset, Somerset, Wiltshire, Hampshire, Derby and Yorkshire. By the end of the eighteenth century the industry had concentrated in the first three of these counties, and by 1780 was employing 140,000 there and over the borders of Huntingdon, Hertfordshire and Oxford. The finest and most expensive lace was made in the West Country – Honiton in Devon and Blandford in Dorset. The trade, however, gradually declined with competition from foreign lace and in some cases from machinery. In Honiton in 1820 only 300 lacemakers remained out of the 21,000 formerly employed.[41] The domestic lace industry appeared to go into a general malaise from 1815 to the 1830s, under those dual threats. While the Nottingham industry had been steadily improving, the pillow lace industry had been 'constantly deteriorating and excluded by the price'. The demand for cheaper products favoured the Nottingham machine-made lace industry. The threat from the handworkers of France was outdone by the machine. 'By the 1830s, the industry had disappeared from a number of places and almost everywhere prices and wages had fallen, and employment had become very intermittent.'[42] But a new wave of demand for lace of all kinds, including in particular hand-made lace, appeared in the 1840s, and the industry gained a new lease of life. Prices and employment reached a high point. After this the industry lasted until 1880.

In addition to lace there was straw plaiting which spread quickly in the borders of Buckinghamshire, Hertfordshire and Bedfordshire. It was also introduced into north Essex in the late eighteenth century, and in 1840 was still keeping women, children and old men busy in the Halstead, Braintree and Barking districts. Straw plaiting had started in the late eighteenth century and spread rapidly just at the time when the decline in wool spinning had left many women unemployed. There were good wages until the end of the Napoleonic Wars.

The introduction of hats from Italy was a temporary setback, and the industry recovered for a time by importing Italian straw. Wages, however, fell to 5s–7s a week and stayed there until the industry finally died in 1870.[43]

The appearance of lace and straw plaiting in southern arable farming regions may have been connected to contemporary agricultural change. As was argued recently, the kind of agricultural changes which took place in the South and the East probably reduced the potential for the participation of women in the agricultural labour force. This was not nearly so marked in western pastoral regions, for such regions did make available some comparatively well paid agricultural and alternative female employment. Because farming activities were reduced for women in the South and the East, a labour force was left to take up lacemaking and straw plaiting. It could just as well be argued, however, that the attractiveness of these industries after the mid-eighteenth century might itself have contributed to the sexual division of labour in agriculture.[44]

There was a rapid increase in glovemaking from the end of the eighteenth century, and this industry found its urban headquarters in Woodstock, Yeovil and Worcester, with out-working villages in Oxfordshire, Somerset and Worcestershire. Worcester and its outlying villages claimed 30,000 workers in the 1820s, Somerset had 20,000, Hereford 3,000. But there was a great decline after Huskisson's withdrawal of the import restrictions on French gloves.[45] By 1832 output at Worcester was one third what it had been in 1825.[46]

Dorset had further cottage industries in string, pack thread, netting, cordage, and ropes with sailcloth and sacking. It was well known, too, for its wire shirt-button industry which in 1793 employed 4,000 in the town and neighbourhood of Shaftesbury. But this industry too bowed out in the 1830s, to competition from horn and pearl buttons.[47]

The decline of the older domestic manufactures in the South of England was thus much more complicated than is generally allowed for. The decline of the older woollen manufacture was of course dominant. Wool, which was Britain's chief home as well as export industry, provided 70 per cent of domestic exports in 1700 and 50 per cent in 1770. The enormous effect of its transfer from the South to Yorkshire, which produced 20 per cent of output in 1700 and 60 per cent in 1800,[48] outweighs all other trends. But still the South did not entirely turn to agriculture – new cottage industries developed on the burial mounds of the old. The reasons for the decline of the old cloth centres cannot lie entirely in a comparative advantage for agriculture in the South, because the appearance of these new industries, limited though they were in extent and long-term success, contradicts that case. The argument of comparative advantage is, in addition, an anachronistic one, because investment decisions were based on a long tradition of mixed portfolios and responses to the trade cycle. Bankruptcy, not comparative advantage, determined the allocation of resources between the industrial and agricultural sectors.[49]

Other regions

Other regions of Britain played an important part in the early phase of industrialization, but failed to sustain their leadership. Cornwall was one such area. Early and rapid growth was based on the tin and copper industries. Because coal was expensive, the area was the first to use the Watt engine, and tin and copper mining and smelting became the basis for one of the most advanced engineering centres of the world. But in the middle of the nineteenth century mining suddenly declined and the region was rapidly transformed into a holiday resort. Shropshire was another such early developer. A technological leader in ironmaking and -using, it fostered brickworks, potteries, glass and chemical works, armament works and engineering plants. But the region was burned out by 1815. The problem here does not appear to have been depletion of mineral resources, but instead failure to use them effectively. Local iron-using industries were not built up, and a decline in the relative wages of ironworkers soon resulted in the emigration of skilled men. Even south Staffordshire and the Black Country failed to stay the course. Rich in coal, iron and water power as well as

generations of skilled metalworkers, it developed some of the
earliest canal networks and glass, engineering and armament
works. But after substantial expansion in heavy industry in
1810–30 the area rapidly declined. The place of North Wales
was also ambiguous. With supplies of coal, slate, iron, lead,
copper and water power it created ironworks, copper-smelting
plants, engineering works, brick- and limeworks. It built up
cotton mills, and woollen-, linen- and ropemaking industries.
Yet even before the depletion of its coal, copper had given out in
the 1820s, and cotton spinning declined by the 1830s. It too
became a recreation area. Derbyshire was rich in lead ore and
water power. It contained a long tradition of domestic textile
work with leading textile innovations including those of
Arkwright, Lombe, Paul, Hargreaves, Cartwright and Strutt. Its
attraction to early cotton entrepreneurs was based on its low
wages and absence of any history of machine-breaking as well as
its proximity to the framework-knitting centres which created
cotton twist. But the cotton industry only lasted fifteen years in
the area before shifting to Lancashire.[50] During this time the
centuries-old leadmining industry also suffered a demise. The
industry's peak output was in the mid-eighteenth century, but
technical changes at the end of the century both displaced the
smaller investor and increased productivity. Trouble was
obvious in the first years of the nineteenth century when the
exhaustion of ore shoots, deeper working and drainage problems led to a reduction in mining output. The industry was in
severe decline by the 1830s.[51]

Abortive development also featured in Ireland, where the
cotton industry had made a brief start in the areas around
Belfast, Dublin and Cork. Cork was already in decline in the
first decade of the nineteenth century as large general manufacturers collapsed following the decline in the Atlantic trade in
provisions. An industry which employed about 80,000 in 1810
succumbed by the 1820s to international commercial crises and
stagnation in the home market. The crisis of 1825–6 caused the
greatest run of failures in the history of the industry, and around
Bandon, where coarse cords were manufactured, a slump in the
mid-1820s followed by the advent of the power loom had a

devastating effect. Bandon decayed, its inhabitants choosing emigration. Other areas responded to the challenge of improvements in flax spinning, and many Belfast cotton spinners opted out of cotton and transferred operations to the linen manufacture.[52]

Cyclical factors

The fortunes of those regions which dropped out of the industrial vanguard in the course of the eighteenth and early nineteenth centuries were partly set by cyclical swings over the period. Industries highly dependent on export markets suffered deeply from the setbacks of the numerous eighteenth-century wars. Deane and Cole saw a definite discontinuity in economic growth in the second quarter of the century. They found a turning point in the rate of growth in total output and incomes as well as in the growth rates of particular industries in the 1740s. This downturn in economic growth was closely connected with the depression in agriculture in the 1730s and 1740s. But this view and the estimates it is based on are now subject to substantial criticism. Recent research conveys a much more optimistic interpretation of growth over the whole century.[53] But the impact of war over the period was probably felt in the index of industrial production, for most eighteenth-century wars had a particularly adverse effect on the home market.[54] Taking the available estimates of output along with contemporary observation, there still seems some evidence for an industrial setback in the years before mid-century.

Complaints of depression in the woollen industry in the 1730s were complemented by an obvious need for export subsidies in the linen manufacture. Innovations in cotton technology – the flying shuttle and roller spinning stood out – took until the 1760s either to spread significantly in the one case or to be perfected in the other. The hosiery industry expanded rapidly in the later seventeenth century and first quarter of the eighteenth, then slackened. Imports of raw silk increased slowly until 1740 then fell to levels lower than at the beginning of the eighteenth century and revived only in the 1750s. The hosiery industry in Nottingham and Leicester fell into widespread poverty in the 1740s and 1750s. Military expenditure during the War of the Austrian Succession boosted the output of Birmingham only slowly. The pottery industry, after undergoing a great transformation in 1690–1720, was in crisis in the late 1750s, and only revived in the third quarter of the eighteenth century. The output of the paper industry, after rising fourfold between 1710 and 1720, stagnated until 1735–45. Domestic consumption of copper and brass fell between 1725 and 1745, exports were booming but prices low. Tin had become almost unsaleable by the late 1740s. The throughput of the English iron industry stood at 26,000 tons in 1625–35 but at only 20–25,000 tons in the 1720s. In Phyllis Deane's words, 'The evidence suggests that the English iron industry in the first half of the eighteenth century was scattered, migratory, intermittent in operation and probably declining.'[55]

This mid-century cyclical industrial malaise may have been enough to set the seal of de-industrialization on some of those areas of former proto-industrial splendour. But it only brought to a head the many underlying institutional problems of the older industrial regions. Long periods of stagnation and at best uncertainty made the prospect of new frontiers all the more appealing.

CHAPTER 6

Domestic Manufacture and Women's Work

For much of the eighteenth century industry was dispersed, both in town and country, in domestic units of production and workshops. I have thus far examined the relationship of industrial change to the wider economy and to the longer-term process of industrialization. Equally important, however, is the microeconomy of the domestic and workshop systems of production and their workforces. Most of the interest of historians has concentrated, I have argued, on the discussion of transitions away from the domestic system towards the factory system. The domestic system has been associated with static technologies, primitive industrial organization and pre-industrial social values. Here I will turn to look at the domestic system in its own right and examine the extent to which this form of industrial organization in all its many varieties contained its own peculiar dynamic. Historians have, furthermore, been interested in the characteristics either of factory labour or of highly skilled artisans; the workforce of domestic industry has occupied only the fringes of historical discussion. This chapter, then, will focus on domestic workers. It will make the case that as household production much pre-industrial manufacture made special use of a female labour force, adopting types of technical change specially suited to the gender and the work patterns of its labour force.

Domestic system and family economy

The special dynamic of the domestic system was created, it has been argued, by the juxtaposition of the traditional peasant economy and the world market. Hans Medick has ascribed the rise of rural cottage industry based on foreign and colonial markets from the sixteenth to the eighteenth centuries to the conjuncture of three conditions: first, population growth and with this the socioeconomic polarization of the rural population; second, the emergence of a world and especially a colonial market; and third, an organizational structure based on the traditional family economy orientated to self-regulating sub-sistence.[1] David Levine has described the system as one of nascent capitalism undermining the traditional social controls which had maintained a demographic equilibrium in peasant societies.[2] The traditional peasant economy had always faced two alternative family strategies: first, to restrict inheritance and to force surplus children out into the world; second, to subdi-vide family holdings among increasing numbers of heirs. The equilibrium of the peasant family economy, before the use of domestic industry, had been maintained via a balance between land and labour. The age of marriage was kept high by the inelastic demand for labour in a pre-industrial economy. The traditional family subsistence economy was one which regulated the intensity of production, the size of the return to labour, and the level of consumption in order to strike a balance between work effort and the satisfaction of family needs. Where a family's priority was to maintain total labour income, any in-crease in population had the effect of driving up subsistence needs and thus forced greater work effort in order to increase the total return to labour. But when subsistence needs could be met more easily (thanks to good harvests or smaller family size) the family would reduce its work effort and put any extra time or economic surplus into material, cultural or ritual consumption.

It is argued that it was the rise of domestic industry with its new world markets which upset the traditional restricted de-mand for labour, and so broke down the old limitations on the age of marriage and the size of the landholding. The result was population increase and land fragmentation. The attractions of the new economic pursuit soon created its own snares. Peasants discovered that they could no longer survive on their agricultural production alone; they needed industrial commodity production to maintain their subsistence. This peasantry became integrated into a larger 'intralocal pool of labourers through which bulky mass-produced goods were transported across the Western Europe commercial world'. These rural industrial workers remained tied to the community, yet they were rootless, for their conditions of life were determined by the international, not the local, economy. Their wages were determined with reference not to local prices, but to the international value of their produce.[3] Their connections with agriculture in many areas became increasingly tenuous, and their independence as peasant producers gave way to proletarian dependence on larger merchant manufacturers. They became more and more subject to commercial and not just harvest fluctuations, and they faced periods of enforced idleness as a result of changes in fashions, interruptions in foreign trade, and an economic cycle moving increasingly to the boom–slump pattern that would come to dominate industrial capitalism. Merchant employers controlled the rural workers' access to markets; against the backdrop of international markets they could force wages down below those customarily acceptable to urban workers, and even below subsistence.

The responses open to such rural workers were either to organize in order to resist such wage reductions or to increase output in order to maintain existing living standards. While the first option was often exercised the conditions of dispersed family production, population pressure and lack of access to alternative markets made such resistance generally less significant and effective than in the case of urban workers. The response of increasing output was one which continued out of the traditional peasant economic motivations. Just as in peasant agriculture increasing population and subsistence needs drove the family to augment its work effort, so now in the family industrial unit increasing competitive pressures drove the family to augment its work effort even in the face of falling prices or wages, in order to maintain a basic family subsistence income. This rural industrial labour force was therefore cheaper than that which could be had in towns. As Hans Medick has put it, the system depended on a self-exploitation of the family through the labour process which was greater than that which could be enforced under the relations of production of either capitalist workshop or factory production. This industrial labour was also traditionally regarded as a by-employment. The connection with the land, however tenuous it might have become, formed the backdrop, and the identification of this industry with the supplementary earnings of wives and children precluded the local customary recognition of rural industry as a primary source of income. As Adam Smith argued, the unspecialized nature of the country workforce made it a source of cheap labour:

> Where a person derives his subsistence from one employment which does not occupy the greater part of his time; in the interest of his leisure he is often willing to work for another for less wages than would otherwise suit the nature of the employment. . . . The produce of such labour comes frequently cheaper to market than would otherwise be suitable to its nature.

And thus his example of stockings which could be knitted more cheaply by hand in Scotland than they could be knitted on the loom.[4] It was the special features of a family-based production unit which under capitalist production paradoxically created the possibilities of forcing wages below subsistence. For the primary goal of maintaining the family together would generate greater work effort in face of falling wages, as the response of the desperate. The traditional social division of labour within the family gave smaller customary accord to the economic activities of women and children, and the rural industry which tapped their labour had made access to a labour force which was cheaper by custom. The system of proto-industrialization thus thrived on cheap, infinitely expandable supplies of labour. It was an economic activity which had transformed demographic conditions, making population increase a major source of labour. But, more significantly, it was also an activity which forced the greater intensity of labour and the greater use of all members of the family, particularly that of the cheap labour of women and children.

David Levine has argued that the ultimate tendency of this

process, particularly population growth, was to a form of industrial involution, for population increase itself also influenced the organization of production. As long as labour was cheap and plentiful there was little incentive to undertake capital investment and to increase productivity. Low wages, in effect, meant that primitive techniques were the most profitable.

This analysis of the special dynamic of domestic industry is stimulating and, indeed, inviting, but it contains some major problems. First, it attempts to marry a demographic analysis of peasant behaviour to a Marxist analysis of the pressures of capitalist competition on manufacturing production. The analysis of peasant behaviour, accepted uncritically by historians of proto-industrialization, is based on the study of the Russian peasantry made in 1910 by A. V. Chayanov.[5] Chayanov treated the peasant farm as the fundamental unit of economy; it was self-defining and self-perpetuating, a homogeneous economy based on the family. Its defining characteristics were the absence of a labour market and the operation of the family farming enterprise by family members. Inequalities of family income and farm size were attributed to family size and family life-cycle. This peasant economy, furthermore, struck a balance between labour and consumption, working to fulfil subsistence needs which varied over time according to family size and life-cycle.

There is little evidence, however, even in the case of pre-Revolutionary Russia, that this peasantry was homogeneous; nor that the rural labour market was insignificant. 'While the labour market as a whole was extremely shallow, participation in it was widespread and unequal.' Wealth, furthermore, depended on land and labour, but the Russian peasants faced sharply falling short-run returns to labour, and the rents they paid for marginal land exceeded its net yield. 'Thus there is no satisfactory answer to the question of how peasant families expand their reproducible assets over the family life-cycle.' Finally, the labour-consumer balance implied that rural society operated according to a law of subsistence, but in reality capitalist pressures and imperatives meant that the individual peasant faced prices and costs set by others.[6]

If the dynamics of the peasant economy are so questionable

even in the case of pre-Revolutionary Russia, they are even more inappropriate to eighteenth-century England. There is increasing evidence that English farmers were dominated neither by land nor by family. 'The family plot in England was to a surprising extent disposable by the man who had control of it.' And there was no necessary relationship between farm size and household size.[7] Eighteenth-century Kirby Lonsdale had features in common with several other English parishes – 'the absence of ties between sons and their father's holdings, geographical mobility, hired labour, saving and thrift, late age at marriage and the movement of girls away from the area. In every respect it is a contrast with a classical peasantry.'[8]

If the peasantry or farming community was not fixed within a narrow framework of family and subsistence, there seems no reason why we should transpose this model to the dynamic of domestic industry, or to largely manufacturing communities. Historians have assumed that domestic industry was largely subsidiary to agriculture, and that the horizons of the proto-industrial craftsman agreed with those of the peasant. But clearly many such workers were landless labourers, participating more or less frequently in rural and urban, agricultural and industrial labour markets.

A model based on family and subsistence, the traditional model of the family economy, is clearly not adequate to the task of analysing the domestic system. For the reality of that system involved the individual and the household, wage labour and family labour, market and custom.

The proto-industrial community still awaits its economic theorist. Its structure and dynamic were dictated neither by family subsistence, nor by capitalist labour markets. Did its way lie somewhere in between, or along an entirely different path?

Women and the labour force

Whatever theory of the domestic system we might ultimately develop, two of its essential elements must be the low wages or return to labour, and the flexibility and potentials for increasing the intensity of labour.

Historians have frequently assumed an early division of labour within the family, with most agricultural labour allocated to men. This would make it apparent that the major source of labour for domestic industry was the women and children of the family. Yet the significance of women's and children's economic contribution is usually passed over in discussions which dwell on family production systems before the factory. Our knowledge here, as a recent historian points out, is still very limited.

> We are thus carried into the realm of women and work and are confronted with a curious paradox. We all *know* that women in pre-industrial society worked. In adolescence and early adulthood, domestic service helped a girl to accumulate the few pounds she needed to constitute a dowry. . . . We know that women worked in hoeing and weeding, in the dairy and the *basse cour* and in domestic industry as well. Sense tells us that in the proto-industrial phase their role was crucial. They were the more numerous sector of the cheap labour force. Yet we have very little detailed modern research bearing on the nature and importance of their labour.[9]

Some of the historians of proto-industry have made the provocative hypothesis that women took on a special significance for the proto-industrial labour force, apart that is from demography. Hans Medick has argued that it was the productive effort of women and children in domestic industry which contributed the necessary share to the family wage, without which the subsistence gap would not have been closed. This labour was not, however, properly compensated, for it did not result in a proportionate increase in income. In fact the decisive marginal work effort of the family remained underpaid. The larger proportion of the labour time of these women went to merchant capitalists in the form of extra profit.[10] And David Levine argues that the restructuring of the domestic economy which occurred with the proletarianization of peasants and craftsmen reordered the division of labour within the family. As well as her normal chores, the wife was now also expected to earn her wages, though direct money transfer did not always occur.[11]

Both these arguments, however, implicitly assume that women had a rather minor role to play in agriculture, and indeed in the urban trades before the spread of proto-industrial manufacture. In fact, they credit proto-industry with a transformation of the division of labour between the sexes. In reality women had a much greater role in the agricultural labour force than is generally assumed.[12] There is also substantial evidence of their importance and sometimes high status in the pre-industrial urban trades. Most of this research has, however, been focused on continental industries.[13] The women of the English urban trades have not been studied since Alice Clark's *Working Life of Women in the Seventeenth Century* (1911).

The employment of women in eighteenth-century proto-industrial occupations may not have been 'novel',[14] but the expansion of these industries and their reliance on low-paid labour did entail much higher proportions of female and child labour. Even in areas of arable cultivation, where the growth of domestic industry might have been restricted through landlord policies, the system of closed parishes and the exigencies of corn husbandry, women's domestic industry flourished. Landlords who needed to restrict the alternatives open to their male agricultural labour forces did not object if women and children took up cottage crafts.[15] The earnings of women and children in such crafts were usually essential to family subsistence.

Such industrial by-employments, depending on time and place, could make the difference between subsistence and destitution, or even provide moderate comfort for households. 'Landless commoners [in eighteenth-century Northamptonshire] supported themselves with fuel, pasture, browse, food and other benefits from commons, with seasonal agricultural labour and with weaving, spinning and carding, woolcombing and stockingmaking, matmaking, shoemaking and lacemaking which enabled the whole family to work.'[16] Casual earnings also derived from the petty productive work within the household carried out by women and children to supplement the primary wage-earning or productive activity of the household. There were

included:

multifarious unco-ordinated and sporadic activities such as gardening, spinning, knitting, straw-plaiting, broom-making, and a whole range of other handicrafts which went forward within them. . . . Many of the women in these households brought in money, although they often had young children . . . the children added little sums too, beginning at six years of age. Trifles of the same kind, mostly derived from knitting undertaken by every household member . . . are detailed in a listing from the village of Corfe Castle in 1790.[17]

Women and children certainly became an attractive pool of cheap labour for merchant manufacturers. Spinning was the archetype of the female domestic craft. Ivy Pinchbeck pointed out that the number of spinners, all women and children, far outnumbered other workers in the textile industries.[18] Eden in the 1790s found that in Essex the majority of women were employed in wool spinning in the towns as well as the countryside.[19] The second most important women's industry was lacemaking. In 1714, Ralph Thoresby found in northeast Bedfordshire, an entirely rural district, ' a low, moist country, abounding with willow, of which are made osier baskets, screens etc. with bobbin lace, seem as the chief manufactures of these parts, and hardly any adult men would have been employed in such work'.[20] Handknitting also occupied women and children over many areas of the country.[21] Women were similarly occupied at a wide range of other rural industries – glovemaking at Woodstock, in Dorset and in the Welsh borders, button-making and straw plaiting in south Bedfordshire and north Hertfordshire, the silk manufacture and metalworking in the West Midlands. The earnings of these women and child workers were not always minimal. In the Staffordshire Potteries in the late eighteenth century the contributions of women and children could more than double family incomes.[22]

The significant economic position which women might hold in a rural agricultural-industrial household is demonstrated

Flax spinning in County Down
(Ulster Folk and Transport Museum)

most clearly in the linen manufacture. Brenda Collins has
broken down the sexual division of labour in the model farming-
weaving family in the pre-famine Irish linen industry to demonstrate that if male weavers formed the linchpin of the manufacturing system, it was the female spinners who were its vital
components.

> Those households without sons of an appropriate age
> employed journeymen or apprentice weavers, because they
> required enough male weavers to engage in domestic in-
> dustry, by which employment is afforded to the females and
> children and the most profitable way of disposing of the flax
> produce, the consumption of potatoes and other produce
> without carrying them to a distant market. . . . It was equally
> important for the farming-weaving household to have the
> appropriate labour supply in the other textile processes. . . .

Most important of all was the need for spinners to provide the yarn for weavers particularly as the ratio of spinning to weaving labour was not on a one to one basis.

The women were in fact more vital than the men to the viability of the linen-manufacturing household for much of the eighteenth century, for linen yarn had a ready sale both within the linen industry itself and as warps in the cotton industry. Households, then, could be formed entirely of women, or it was not uncommon for cottiers 'who have families of industrious females to take larger portions of flax ground'. Weavers, on the other hand, either had to have female spinners in their households or they had to buy their yarn on the market. And 'as the extension of the industry led to geographical specialization, the trend of spinning households entering into a direct cash nexus independent of a familial relationship with male weavers, became more important'. The yarn produced in such households in the northwest of Ireland 'was sold to rural weavers in southeast Ulster.[23]

The skewed markets of the eighteenth-century linen industry only reinforced a more general tendency to the horizontal division of the production process between preparatory and spinning processes on the one hand, and weaving on the other. Linen cloth was best woven with bleached yarn, and the bleaching process took several months. Few small clothiers could afford to hold several months' stock of yarn, and hence the tendency to household division of the manufacturing process.

Female technologies and women's skills

The constraints imposed on the expansion of the textile industries by the supply of spun yarn is the common explanation for the search for spinning machinery and the rise of the first spinning factories. Yet the importance and scarcity of spinning labour implied in these explanations were not reflected in wages: spinning remained an entirely female trade at least until the advent of the larger spinning jennies, and the women who practised it right across the country were invariably among the lowest paid of workers. Eden found that the earnings of female domestic spinners in Essex, Norfolk, Oxfordshire, Leicestershire and Yorkshire ranged from 3d–8d a day or 1s 6d–3s a week. The women who worked in three Yorkshire cotton mills earned 4s–5s a week while those employed in the Birmingham toy trades could gain 7s–10s a week. This range of payment still placed the wages of most women workers well below those of the lowest paid male labour. Male agricultural labourers earned about 8s a week at the time.[24] The poverty of female woollen spinners was best expressed by Julia Mann who referred to them as 'an unorganized mass of sweated labour'.[25] And Arthur Young attacked the Norwich master manufacturers in 1788 for the wretched state of their spinners:

> The suffering of thousands of wretched individuals, willing to work, but starving from their ill requited labour: of whole families of honest, industrious children offering their little hands to the wheel, and asking bread of the helpless mother, unable through this *well regulated* manufacture to give it them.[26]

And yet such low wages were subject to yet greater reduction. Eden in 1790 found that the war had led to a depression in wages in the Halifax region, and 'many poor women who earned a bare subsistence spinning are now in a very wretched condition'.[27]

The low cost of this women's spinning labour also made it viable to continue using the primitive distaff long after the introduction of spinning wheels and even spinning jennies. This was partly because it was possible for a time to spin much finer yarn on the distaff than on the wheel or jenny, but the major reason for its survival was that the distaff could tap labour not otherwise in use – that of feeble old women and young children, and the hands of women not otherwise in use when walking, talking, tending animals or watching over children.

Spinning with distaff and spindle
(*The Useful Arts and Manufactures of Great Britain*, n.d.)

The flax spinning wheel
(*The Useful Arts and Manufactures of Great Britain*, n.d.)

At the end of the eighteenth century Eden found the distaff still widely used in Scotland and noted that 'one rarely met with an old woman in the north of Scotland, that is not otherwise employed, but who has got a distaff stuck in her girdle and a spindle at her hand'.[28]   Alice Clark cited an eighteenth-century commentator who discussed the choice between the two techniques:

There are, to speed their labour, who prefer wheels double
spol'd, which yield to either hand. A sev'ral line; and many, yet
adhere to this ancient distaff, at the bosom fix'd casting the
whirling spindle as they walk.[29]

Spinning on the distaff occupied hands otherwise unoccupied, or
left the other parts of a woman's body free for yet more work, such
as was found by Hugh Miller in the Scottish Highlands even as
late as 1823:

Here as in all semi-barbarous countries, is the woman seen to
be regarded rather the drudge than the companion to the man.
The husband turns up the land and sows it – the wife conveys
the manure to it in a creel, tends the corn, reaps it, hoes the
potatoes, digs them up, carries the whole home on her back,
when bearing the creel she is also engaged with spinning with
the distaff . . .[30]

Machinery and women's work

When the spinning jenny was introduced it met with great resistance
in the woollen industry of the South and East because of the greater
importance there of handspinning to poor women. The more
limited resistance it met in the North resulted only from the
existence there of a more important alternative – the spinning of
worsted, for which the jenny was not used.[31] Indeed, in the worsted
manufacture the one threadwheel was common until the end of the
century and later.[32] A magistrate in Somerset in 1790 described how
he was called in by two manufacturers to protect their property

from the Depredations of a lawless Banditi of colliers and their
wives, for the wives had lost their work to spinning engines . . .
they advanced at first with much Insolence, avowing their
intention of cutting to pieces the Machine lately introduced in
the woollen manufacture; which they suppose, if generally
adopted, will lessen the demand for manual labour. The
women became clamorous. The men were more open to
conviction and after some Expostulation were induced to
desist from their purpose and return peaceably home.[33]

Spinning jenny
(C. Aspin and S. D. Chapman, *James Hargreaves and the Spinning Jenny*, 1962)

The dispute over the introduction of spinning jennies was a long and contentious one. The jenny substantially increased the productivity of women spinners and brought higher wages, though the extent of wage differences was a matter for debate. In the 1760s it was said that handspinners earned 3d–5d a day and jenny spinners 1s–1s 3d. But the opponents of the machine in 1780 estimated the jenny spinners' wage at 8d–1s a day; while the advocates claimed 2s–2s 6d for jenny spinners and only 3d–4d a day for handspinners. 'For some years a good spinner had been able to get as much or more than a weaver.'[34] Most of the disputes, however, were over the large jenny used in jenny factories rather than at home. Baines wrote of the desperate efforts of workpeople in 1779 to put down the machine. 'The mob scoured the county for several miles around Blackburn, demolishing jennies, carding engines and every machine turned by water or horses, sparing only those with less than 20 spindles.[35] And Wadsworth and Mann have described part of the country as in a state of 'guerrilla warfare' in the autumn of 1779; the 'spinners' interests were those of every working-class family'.[36] Similar disputes over the taking away of women's labour with the introduction of machinery had arisen even earlier in the silk manufacture. In the 1730s new looms, 'like ribbon looms that could do as much as 8 to 10 hand workers and as cheaply as 6 to 8' were introduced to weave narrow strips of material for buttons. This dispossessed the buttonworking section of the population of their needlework, 'ancient and decrepit men and women, and children'. And in 1737, the women of Macclesfield 'rose in a mob and burnt some looms, and when their leaders were arrested, released them from prison'.[37]

The mechanization that came with industrialization was supposed to have destroyed many of the widespread family-based women's trades. Eric Richards has argued that with industrialization the gradual loss in employment in traditional lines probably exceeded the emergence of new opportunities. The introduction of new technology increased structural unemployment for women.[38] And Clapham argued much earlier that spinning machinery, knitting and lockworking implements had left women's hands idle and family earnings curtailed all over the countryside in an age of hunger and high prices.[39] Jones has also pointed out that mechanization drove many of the handicraft districts into industrial oblivion, cutting deeply into the base of 'mother and daughter power' of the South and East.[40]

There was a major decline in cottage industries after 1815. 'The decline of female spinning in particular, probably most marked after 1800 . . . may have aggravated female unemployment and depressed familial income.'[41] But that decline was not final, for to argue thus is to mistake the process of capitalist expansion. The decline of some cottage manufactures could open the way for other, more degraded, employment, and not just domestic service. Technological change and factory production formed but one part of the route to industrialization.

The search for ways of using more but cheaper labour more intensively was equally a way of increasing profits and expanding capital. In Hobsbawm's words, 'The obvious way of industrial expansion in the eighteenth century was not to construct factories, but to extend the so-called domestic system.'[42]

Women and labour-intensive technologies

Landes referred to the equal significance of 'capital widening' to 'capital deepening',[43] and Sidney Pollard called the labour-using proclivity of capitalist expansion 'the inner colonization of labour'.[44] In fact, cheap women's labour, driven in the process of mechanization to even lower wage levels, was not simply left unemployed. It became a source of new cheaper labour in the late eighteenth and early nineteenth centuries for new rural industries in lace, strawplait manufacture, glovemaking and shirt-buttonmaking, and for the new urban sweated trades which arose and flourished from the 1830s onwards.[45] This cheap women's labour was a lucrative source of profit not to be by-passed by manufacturers ready to launch new labour-intensive industries. Women's cheap labour in combination with hand or intermediate techniques long continued to be chosen as an alternative to mechanization. And though mechanization might threaten such labour, new use was usually found for it. When machine spinning came to the linen industry, women's labour was transferred to a lighter and simpler weaving industry. When new techniques were invented which still required some skilled labour, alternative labour-intensive techniques were used or even invented to tap a large cheap female labour force.

The adaptive use of the employment structure to overcome problems of skill or technique was a way of avoiding the necessity of introducing whole new technologies. Perhaps the best example of this development of older as well as new technologies geared to particular labour forces is provided by calico printing. This was an industry which developed labour-intensive technologies along with an advanced division of labour in order to tap a female labour force. Fears of competition from inexpensive and labour-intensive Oriental printed fabrics stimulated four technical innovations. The first was 'picotage' or the

Calico printing – 'pencilling' and block printing
(*Supplement* to *The New and Universal Dictionary of Arts and Sciences*, 1754)

patterning of printing blocks with pins or studs tapped into the blocks. This was delicate work, for one large block contained 63,000 pins, but it was a job done by women who earned 12s–14s a week after their apprenticeship. Another labour-intensive process introduced at the time was 'pencilling' or the hand-painting of patterns directly on to the cloth. This was performed by women who worked in long terraces of cottage-like work-shops under the superintendence of 'mistresses'.

In the shop each woman had her piece suspended before her with a supply of hair pencils of different degrees of fineness according to the size of the object . . . to be touched, and containing colour . . . according to the pattern required . . . a good workwoman might earn £2.00 a week, though it was likely most earned a lot less.

Block printing
(*The Useful Arts and Manufactures of Great Britain*, n.d.)

The style of patterns changed little from year to year. This laborious work was done by women, and so was regarded as an unskilled process which by-passed the employment of the highly paid craftsmen who engraved and used wooden printing blocks and, after 1760, copper plates. Copper plate printing, introduced in 1760, followed by roller printing in 1785, constituted the real technical improvements of the industry, but they required the use of highly organized and highly paid 'gentlemen journeymen', so that such manufacturers as Peel by-passed these and instead organized 'protofactories' using elementary labour-intensive techniques and extensive division of labour, along with special training and disciplining of workers. The scarcity, together with the very high status, of skilled calico printers was the main stimulus behind the attempts by entrepreneurs to look for alternative methods of production on which they could employ low-paid women and girls.[46]
Another important new industry which tapped the labour of

Cylinder printing
(*The Useful Arts and Manufactures of Great Britain*, n.d.)

women and children in declining woollen regions was the silk manufacture. Before and after mechanization silk throwing was carried out by the cheapest labour available – that of women and children. The introduction of silk-throwing machinery simply reproduced the hand procedures on a larger scale. Large supplies of female and juvenile labour were still used to tie threads and assist with winding. Where there were large supplies of such labour, manually operated winding and throwing mills continued to be used long after water- and steam-powered silk mills had appeared. This was particularly so in East Anglia, and the mills which subsequently grew up in the area remained disproportionately dependent on the labour of young girls, and paid disproportionately low wages.[47]

Sexual divisions within technologies
A sexual division of labour between trades and within branches of trades, in fact, complemented a specific sexual division of

labour within work processes and technologies. To some extent women were simply confined to the use of more labour-intensive and less efficient techniques where skilled workers were able to restrict entry. Although women were traditionally the spinners, they were only allowed to continue their work at distaff, wheel and jenny after the introduction of the spinning mule, for mulespinning was successfully appropriated from the outset as the preserve of male workers.

Handspinning of woollens and worsteds was the major industrial employment of women throughout the eighteenth century. Women also dominated the domestic jenny-spinning stage of the cotton industry, and the female spinners formed the backbone of the linen industry, at least until the widespread introduction of Arkwright frames to produce an adequate cotton and, later, linen warp. The productivity of the female domestic linen spinner was doubled by the introduction of the two-handed spinning wheel in 1770, just as that of the cotton spinner was raised several times by the introduction of the jenny. But many women still continued on the older technologies, for the two-handed spinning wheel was not widely distributed until the end of the eighteenth century.

A skilled activity such as mulespinning was so defined from its earliest days in the domestic system, and was assumed to require the use of male labour. Early requirements like strength, skilled spinning, building maintenance and repair skills, as well as a certain amount of capital, accounted for the 'maleness' of the occupation. But the early application of water power in the 1790s which eliminated the need for special strength, and subsequently the introduction of the self-actor in the 1830s made no difference to the sexual division of labour. Women were excluded from a technique which was defined simultaneously as male and skilled. Female cotton weavers in the eighteenth century were almost invariably found outside the urban smallware manufacture which also used the Dutch loom. They worked at the ordinary handloom in the more loosely organized country branches of the check, linen and fustian manufacture. Dutch looms in fact tended to be grouped in the workshops of a superior class of master weavers, and workers were regarded as skilled, their seven-year apprenticeship leading on to entry into the small master class. Some women did weave on the Dutch loom, but these were almost invariably widows of former smallware weavers. The most common case is exemplified by the daughter of a fustian weaver who had three looms. At twenty she married a weaver of checks and stuffs for women's gowns, and they worked together on two looms. As the children grew up, more looms were added until they had five. After her husband's death and until she was seventy she kept three looms.[48]

Cases like this confirm a gender specification of technological development in the eighteenth century very similar to that found in the agricultural systems of developing countries today. Esther Boserup found that modern agricultural methods neglected the female agricultural labour force and caused men to monopolize the new equipment and methods. Women were left to perform the manual tasks while men used the efficient equipment. The result was an increase in male labour productivity while women's stayed static. 'Such a development has the considerable effect of enhancing the prestige of men and lowering the status of women.'

It is the men who do the modern things; men spread the fertilizer in the fields, women spread manure. Men ride the bicycles and drive the lorries while women carry head loads. Men represent modern farming in the village; women the old drudgery.[49]

Skill definitions
But there is also an important sense in which the jobs and techniques to which women were confined should not just be written off as unskilled: the special attributes they brought to their work processes certainly generated increases in productivity, and the very definitions of skilled and unskilled labour have at their root social and gender distinctions of far greater significance than any technical attribute. As feminists have pointed out, women are particularly sought out by employers for their 'nimble fingers', and their powers of concentration on

tedious, laborious processes, as well as for their docility and their cheapness. These nimble fingers by repute derive from a long but totally unacknowledged training in household arts and needlework. It was these that formed the background to the acquisition of the knacks, the deftness and the special application with which women worked. But these 'female characteristics' were never regarded as skilled in their own right.[50] In the eighteenth century women were particularly sought out for work characterized by delicacy and repetition – as block pinners and pencillers in calico printing, for the intricate work of stamping and piercing in the button manufacture, for painting and decorating in the toy trades and the potteries, and for stove and polishing work in the japanning trades. Such skilled women were sought out, but still undervalued. In Wedgwood's London workrooms in the early 1770s a skilled female flower painter earned 3s 6d a day, or two-thirds the top rate for skilled male painters who received 5s 6d a day.[51]

Skill has traditionally been associated with masculine virtues. Male skills created a solidarity which extended beyond the workplace.[52] It was also the case that men defined their work as skilled and that of women as unskilled, for reasons of social status above all other reasons. As Phillips and Taylor have described it, immigrant men in late nineteenth-century America entered the female-dominated clothing trade because they were excluded as immigrants from traditional male trades. Facing such social exclusion, they needed to establish and maintain some social status within their own communities and families. Thus they termed the processes they worked on as skilled; those of their women folk as unskilled. Men's struggles to maintain their skilled priority in the workforce against machinery and against the encroachment of unskilled women was thus an important part of their effort to maintain their social status within the community and within their families. This familial, customary and social division of labour then took priority over and largely determined the technical division of labour.[53] The sexual division of labour therefore was part of the social hierarchy established among activities. As Maurice Godelier has put it, in primitive societies hunting 'is often more highly regarded than gathering or agriculture. In societies where men dominate, women's tasks are considered inferior only because they have been consigned to women.' In other words, the division of labour is an effect of the social hierarchy and not its cause.[54]

The discussion of the relationship of family and community to skills within domestic manufacture leads on to two issues. In the first place there is the difficult question whether, even if women's labour was so cheap and undervalued, the prospect of new employment opportunities in the domestic industries enhanced women's status within the family. Secondly, we must ask whether the impact of custom and community in the workplace differed in proto-industrial production from that in the organized workshop trades: did the spread of proto-industrialization affect the values associated with the workshop trades?

Family, status and training

Hans Medick has argued that proto-industrial production entailed a change in the division of labour within the nuclear family. Domestic manufacture placed a premium on early marriage and high fertility, as well as the greatest possible work capacity and technical skill of both marriage partners. He argues that 'women were in the vanguard of peasant household industries' and that with the increasing importance of these industries in providing for the family subsistence the men were drawn in from the fields and back to the household. The positive result was a more flexible allocation of role responsibilities for family members than had been the case in peasant families. The old matrimonial control of the young through the allocation of land was broken down by new employment opportunities as well as by land fragmentation. Youths could therefore marry sooner and start their own households; indeed, they had a positive incentive to do so because the opportunities for maximizing their income depended on their capacity for work as well as the number of child labourers they would produce. The adult proto-industrial worker was unable to exist on his own; his productive output depended on the cooperation of his entire family.[55]

As we have argued already, there is evidence that women did a wide range of work in pre-industrial agriculture and manufacture, and it is difficult to sustain an argument that the advent of proto-industrial manufacture transformed the household division of labour. Caution must rule our statements about the sexual division of labour, for our evidence is very slender for the eighteenth century as well as for the sixteenth and seventeenth:

> We still know very little about sexual divisions within the households of the majority of the rural population, about the different activities and concerns of women and men, old and young, about how socialization took place.[56]

Households, above all, were complex and varied. Though nuclear families predominated at any one time, it was frequently the case that the household production unit would require additional labour.[57] Where the age and sex ratio of the family unit did not coincide with that of the production unit, journeymen, apprentices or relations might be taken into the household to back up or extend the productive unit. In the Irish linen industry, the need for female spinners could be met by the 'importation of suitable labour into the household'.[58] Households were also extended in times of falling wages, for the reaction of the domestic producer was to increase output. In the framework-knitting industry as wages fell there was recourse to coresidence rather than reversion to a higher marriage age. In nineteenth-century Shepshed, many framework-knitting households contained resident kin or were shared between two families.[59]

Female households
Another option allowing production processes to be divided and use their own markets (as in the case of spun yarn) was to create separate households of young women or of women and children. The premium on the labour of young women appeared clearly in the Irish linen industry, where such households were common

in the northwest of Ireland. It was also clear from the complaints of contemporary moralists and economists. Anderson declaimed against domestic industry because the money paid for the making up of manufactures would flow into the hands of the lowest ranks of the people, 'often into those of women and children; who becoming giddy and vain, usually lay out the greatest part of the money thus gained, in buying fine clothes and other gawdy gewgaws that catch their idle fancy'.[60] The wool spinners of Bradford Manor were similarly condemned by the Court Leet in 1687 for asserting their independence: 'Whereat many young women, healthful and strong, combine and agree to cot and live together without government, and refuse to work in time of harvest and give great occasion for lewdness.'[61]

Status

If we compare the situation of women in the new manufacturing households to their former position within the peasant family economy, their status *may* have improved, depending on the agricultural and industrial conditions which varied over time and place. But where they remained within the family production unit, their labour was still far cheaper than it would have been, and indeed was, in workshops or early factories. Ivy Pinchbeck in fact blames the domestic system for weakening women's former position. The tradition of low wages established within the family industrial unit contributed to the subsequent low wage levels offered to women who did enter the factory. On the one hand, as the workshop trades faced increasing competition from the sixteenth century onwards, they started to exclude women from apprenticeship. On the other hand, the availability of domestic industry within the family economy meant that girls could be usefully employed at home, and this prevented many of them being put out to apprenticeship.[62]

The tradition of service in another household during adolescence, which was an important aspect of sixteenth- and seventeenth-century English social structure, appears to have declined in the eighteenth. There were some communities where, in the earlier period, service was regarded as a 'bleak alternative' for girls who preferred to stay at home until marriage. Some female servants were 'isolated' and 'powerless' compared to the stronger social position of women able to live in the households of parents and kin.[63] But it is also the case that, though such women may have been able to summon the support of their families as compared to more isolated servants in a community, their social position as workers and individuals must have been very constrained by patriarchal authority at home.[64] Such social tensions operated at the level of the household and of the community, and preferences must have been partly con-ditioned by wider attitudes to women in the community and in the society at large. Social reasons such as these, combined with the availability of more household employment in cottage manu-factures in the eighteenth century, may have contributed to a decline in the practice of sending adolescent girls as well as boys into apprenticeship and service.

Training

There was also, however, the important role in organizing and training which women carried out in domestic production. Children were integral to the production processes of a great many pre-industrial and proto-industrial manufactures. They were employed from as young as six years of age in the workshop trades and at home. Highly skilled calico printers using both traditional and more advanced techniques relied on children as assistants, and very young girls were widely employed in the bleaching fields. The traditional drawboy looms always relied on a child assistant.[65] In household manufactures their labour was taken for granted. Before the advent of factory spinning 'child spinners were trained by females'. In cotton and in wool, 'the mother was responsible for all the preparatory processes and the training and setting to work of the children'.[66] Children were taught to carry out the same types of industrial activities as women; they formed an equally important part of the proto-industrial labour force and they were invariably trained by women. Radcliffe, the inventor of the dressing frame, re-membered how my mother taught me [while too young to weave] to earn my bread by carding and spinning cotton, winding linen or cotton weft for my father and elder brothers at the loom, until I became of sufficient age and strength for my father to put me into a loom.[67]

And it was not only the spinning and preparatory processes which women taught but also the weaving. The Hammonds pointed out that women became weavers in considerable numbers at the end of the eighteenth century. Between 1797 and 1799, when there was a great scarcity of Spanish wool, employment was bad and many men enlisted. Large supplies of wool entered soon afterwards, and women filled the men's places. One employer in Freshford, Somerset, had as many women working for him as men. At Bradford in Wiltshire at least two-fifths of the weavers were women. But little was heard of these women afterwards, and the female weavers left in the woollen industry of the Southwest were by 1840 employed only in the lighter branches of the industry at low rates of pay.[68]

Women trained and supervised the younger members of the family production unit; they passed on 'skills' to the next generation of the industrial workforce and they cared for their children, all as part of one process. Children were involved in the productive activities that women carried out anyway. Even very young children were taught to wind thread and clean wool. George Jacob Holyoake's mother was a self-employed managing mistress of a horn button workshop attached to their house, and she simultaneously made time to raise her family.[69] The care of children was thus integrated into women's productive activity. The labour, management and training roles of women within the family production unit were all highly significant, but the value placed on them was generally small. Still, the intensity of their labour was shaped by the need to fill the gap between indigence and subsistence, however wide that gap threatened to become.

The low status and value of this women's work, in spite of its acknowledged necessity and significance to the cash earnings of the household, must be largely explained by these workers'

continued social subordination within the family. What appears
to have happened is that with the rise of domestic industry
women's low-paid cash-earning activities became increasingly
associated with household duties. Unlike older peasant and
artisan families, where late marriage after a period of service or
apprenticeship away from the home prevailed, girls now worked
in their parents' home until marriage at an early age. They then
set up their own production unit within a new family setting and
had more children earlier because of the premium on child
labour. Their industrial production therefore became inter-
twined with household formation and what we now call house-
work. There was no division between their cash-earning
activities and household duties. The family setting also affected
the training of girls. Although in many of the textile industries
boys and girls were both brought up to assist in all branches of
the trade, girls most often combined household chores with
casual industrial employment. It was precisely these household
chores, especially needlework, however primitive they might
be,[70] which gave women the special dexterity and diligence they
ultimately brought to the production process. But more
significantly, this combination of activities also resulted in a very
irregular training for women, and it was in the training process,
and the customs and conventions associated with it, that entry to
a trade was controlled and skills defined. However necessary
and significant the labour of women in the domestic industries,
the control of these industries went to men, while women were
relegated to subordinate positions. It is also likely that this
subordination of women in proto-industrial production affected
the positions of their sisters in the apprenticed workshop trades.

CHAPTER 7

Custom and Community in Domestic Manufacture and the Trades

If the labour forces and technologies of many domestic industries developed along lines which were gender-specific, so
too did the custom and community networks formed among
workers in their places of production and in their wider cultural
and social activities. A striking difference appears to have existed between the cultural and community basis of rural or
family-based manufacture and that of the workshop trades. This
difference grew wider as women came to be excluded from the
workshop trades, or at least organized into separate trade
societies from the men. The skilled workers of the artisan trades
appealed to a corporate, collectivist and solidarist idiom. They
regarded their skilled trades as 'moral communities' in which
their art was a source of honour. Journeymen fought their
masters in the eighteenth and early nineteenth centuries to
prevent the breakdown of their moral communities before the
onset of competitive individualism. This entailed the enforcement of apprenticeship and the preservation of the 'old ex-
clusiveness of the freedom'.[1] Artisans were well aware that the
breakdown of the trade communities into a collection of equal
individuals through *laissez faire* would actively entail a division
between property owners and the propertyless. The 'old ex-
clusiveness of the freedom' gave the artisan his independence
and thus his 'capability of supporting himself and his family at a
proper standard without recourse to charity or the poor law'.[2] As
Malcolmson has argued, such artisans rejected the new narrow
definition of their rights that considered only their freedom to
sell their labour on the open market. They had a morally
informed conception of their freedom, involving high respect for
'independent as opposed to enslaved labour'.[3] Prothero has
written that this independence meant rights to a decent wage,

avoiding the humiliation of pauperism, and exercising some degree of control over the work processes. Protecting their independence involved association outside the workplace in ritual, custom and the public house.[4]

Artisan organizations earlier in the eighteenth century had generally included women workers where applicable in the mixed trades. The term 'journeyman' frequently covered both sexes. In Manchester in the 1740s there was a mixed society of smallware weavers, and in 1788 an informal union of female wool spinners called the sisterhood stirred the men to riot over pauper labour and machinery.[5] But from the later eighteenth century women were increasingly excluded from the organizations if not from the trades. In 1769 the Spitalfields silk weavers excluded women from higher paid work, and in 1779 journeymen bookbinders excluded women from their union. The Stockport Hat Makers' Society laid down rules in 1808, that all women were to be struck against, one shop at a time, until all had been removed, and the Cotton Spinners' Union of 1829 specifically excluded women. The handloom weavers consistently refused to admit women to their unions and in 1834 the London tailors struck work to drive women from the trade.[6] The women did come to organize on their own, but combinations that had once been sexually integrated were being replaced by sex-segregated ones. The increasing tension between men and women with the debasement of craft skills revealed to many women workers that 'the men are as bad as their masters'.[7] Not only did workers' organizations become increasingly segmented, but the language of artisan institutions and the perception of skill itself became increasingly identified with masculinity.

Community ties and outworkers

The custom and community which impinged upon the work unit of domestic industry, particularly on the women, was one which moved in quite different directions, out to a plebeian culture based on community ties between families and neighbours, not on the ties established between workers in a journeymen's association. The corporate culture of urban workers contained substantial differences from the forms of organization of domestic workers in rural communities. There was not, however, that polarity between the highly organized urban trades and the totally disorganized and dispersed rural manufactures which historians have claimed to be the major reason for the shift of the centre of production from the town to the country. Rather the economic and social motivations of a family-based workforce interacted differently with the local community and rural custom.

There is certainly ample evidence of a very high degree of organization among many rural workers. Dispersed production and a workforce scattered over many parishes did not prevent the cotton check weavers from organizing. They simply devolved their organization. While maintaining a central 'box' in Manchester, the weavers essentially organized themselves through a series of local 'boxes' in the parishes. Strict rules were maintained, as for instance that no member could take more than two apprentices apart from his own children. The employers' attempt to break the union in 1758 led to a great strike of thousands of check weavers for four months.[8] The high levels of organization maintained among rural weavers and framework knitters in the Luddite outbreaks and later by the country weavers in the machinery riots of 1826 are proverbial.[9] As E. P. Thompson puts it,

> In Nottingham and the West Riding in particular, the strength of the Luddites was in small industrial villages where every man was known to his neighbours and bound in the same close kinship-network. The sanction of an oath would have been terrible enough to a superstitious-minded people; but the sanction of the community was even stronger.[10]

To a large extent, too, 'the values of the domestic worker were also the values of the society in which he lived'. Some regions experienced a smooth transition to the factory system, but other communities in the East and the West Midlands supported the resistance of their framework knitters and silk weavers to the advance of machinery. The decline of industry in the South of England can also be partly attributed to more forceful workers' resistance to mechanization. The language of protest among domestic workers against exploitation was 'at least as violent and not less justified than that used against factory masters in the following centuries'.[11]

The high levels of corporate solidarity of workers in the South of England are frequently contrasted to the lower levels of organization in the North. Jones appeals to the socially more uniform small weaver communities of the North as opposed to the sharp polarity between master and man in the South.[12] John Rule underlines the argument that the lack of separation of small domestic clothiers from their journeymen in the West Riding of Yorkshire is held to have produced an industrial context inappropriate to a distinct labour interest; he argues that this was particularly true in woollen production, but less so in worsted.[13] None has put the case more forcefully than Josiah Tucker did in 1757.

> One Person, with a great Stock and large Credit, buys the Wool, pays for the Spinning, Weaving, Milling, Dying, Shearing, Dressing, etc. etc. That is, he is the Master of the Whole Manufacture from first to last, and perhaps employs a thousand Persons under him. This is the Clothier, whom all the Rest are to look upon as their Paymaster. But will they not also sometimes look upon him as their Tyrant? And as great Numbers of them work together in the same Shop, will they not have it the more in their Power to vitiate and corrupt each other, to cabal and associate against their Masters and to break out into Mobs and Riots upon every little Occasion? . . . Besides, as the Master is placed so high above the Condition of the Journeyman, both their Conditions approach much nearer to that of a Planter and Slave in our American Colonies, than might be expected in such a Country as England: and the Vices and Tempers belonging to each Condition are of the same Kind, only in an inferior Degree. The Master, for Example, however well-disposed in himself, is naturally tempted by his Situation to be proud and over-bearing, to

consider his People as the Scum of the Earth, whom he has a Right to squeeze whenever he can; because they ought to be kept low, and not to rise up in Competition with their Superiors. The Journeymen on the contrary, are equally tempted by their Situation, to envy the high Station, and superior Fortunes of their Masters, and to envy them the more, in Proportion as they find themselves deprived of the Hopes of advancing themselves to the same Degree by any Stretch of Industry, or superior Skill. Hence their Self-Love takes a wrong Turn, destructive to themselves, and others. They think it no Crime to get as much Wages, and to do as little for it as they possibly can, to lie and cheat, and do any other bad Thing; provided it is only against their Master, whom they look upon as their common Enemy, with whom no Faith is to be Kept.[14]

Tucker goes on to imply a lack of organization among the rural workers and small clothiers of the North, particularly in Yorkshire: yet it was here that the first cooperative mill ventures were organized, and that only under a trust deed which gave partners no security against each other and no legal redress. As Pat Hudson has argued, 'Much of the success of these ventures thus depended on the mutual trust and cooperation of the clothier communities.'[15] This community cooperation was of long standing and multidimensional. Different villages produced different fabrics or designs and attended different markets. There was 'a strong sense of local allegiance within the neighbourhoods' manifest in the extent of private charity among themselves and the number of 'secret orders, sick clubs and funeral briefs'. In Pudsey 'the clothiers would watch and nurse each other's families when sick, and borrow and lend almost anything in the house'.[16]

In this case it was the agrarian environment and the long traditions of old manorial, communal agriculture which lay behind the close-knit structures of the clothier communities. There were clearly, then, several different types of community which might bind workers together – the corporate basis of guild or trade union organization was not the only form.

Manufacture and the Economy

Community and neighbourhood clearly mattered to rural domestic workers, dispersed though they may have been. Community solidarity formed the vital foundation for the high levels of organization among country workers not only in industrial disputes, but in food riots and enclosure protests. These were bonds formed, furthermore, between agricultural and industrial, urban and rural workers. Workers from the town sometimes came out to the country to level enclosures, and industrial workers squatting on the commons were leading protesters against enclosure.[17]

> Almost without exception, the enclosure of waste was opposed everywhere, and fears of conversion to pasture may have been almost as provocative, but both wastes and conversion were more common and met more resistance where open fields and rural industry coincided.[18]

It was, in addition, industrial workers in rural industrial communities and small market towns who, in the main, led the food riots, and the decline of food rioting in some areas of southern England has been attributed to the decline of the industrial communities themselves: 'The ability of workers from textile communities to intervene collectively in the marketing of food had been related to the strength of solidarity in those communities.' And historians have frequently remarked on the organization and ritual behind these food protests.[19]

Women's networks and the organization of production

High proportions of these domestic industrial workers were women. We can ask to what extent their community solidarity was founded on bonds formed by women and among women. And furthermore, to what extent did these women's bonds and networks affect the organization of work, or the structure of the domestic system itself? These are big questions which we can by no means pretend to answer with the very limited information and sources we have on eighteenth-century English communities and manufacturing. But what we can do is to consider some hypotheses about the relations between work and community, and their special relevance for women workers and household industries. Such hypotheses may at least invite us to re-examine the limited sources we have in new ways.

Historians working on late nineteenth- and early twentieth-century communities have recently turned to the networks formed among women in working-class neighbourhoods, and in some cases the impact of these on workplace organization.[20] They have argued, less convincingly, that women viewed themselves in terms of their relationships rather than their occupations,[21] and because of this they were especially subject to, and also wielded, a potent community control – gossip.[22] In sixteenth- and seventeenth-century Ryton, reputation, gossip and social power were intertwined: 'The older women . . . who had accumulated information over the years, had considerable power to use what they knew, either to damage a reputation or to influence the decisions of the men of the household. . .'[23]

It does seem plausible that such community networks and controls, particularly in rural areas, gave corporate identity to household manufacturing units. The family or household economy would not, then, be an autonomous unit, but a part of the cooperative and collective networks between households in a village.[24] These networks were not, on the whole, based on kin, for even in the sixteenth and seventeenth centuries 'there was a general cultural preference for a nuclear family household as the usual residential arrangement'. They were based on neighbourhood, and there is substantial evidence of extensive lending and borrowing of money between neighbours from the sixteenth to the eighteenth centuries.[25] Some of these networks would be enforced in the cooperative contexts in which some parts of women's household and industrial work took place. The help which women gave each other in childbirth and illness, in childcare and the collective networks based on local putting-out arrangements, fairs and the marketplace itself – all formed strong and vital community bonds. Women's significance in this sphere was also indicative of their role in local custom and protest. It was the women who led the food riots, organized the gleaning, mobbed the poor law officials and dominated such rites as lifting or heaving. Malcolmson cites some telling examples of the force of such female cooperation.

> In June 1753 at Taunton, Somerset, several hundred women assembled in a body to destroy the weir belonging to several grist mills, and thereby prevent corn from being ground at the mills (the men, it was said, 'stood as spectators, giving the women many Huzza's and commendations for their Dexterity in the Work they were about'). . . . On a market day at Exeter in April 1757, some Farmers demanded 11s per Bushel for wheat, and were arguing among themselves to bring it to 15s and then make a stand. However, some of the Townsmen getting wind of this plot sent their Wives in great Numbers to Market, resolving to give no more that 6s per Bushel, and, if they would not sell it at that Price, to take it by Force; and such Wives, as did not stand by this Agreement, were to be well flogg'd by their Comrades. Having thus determined, they marched to the Corn-Market, and harangued the Farmers in such a Manner, that they lowered their Price to 8s 6d. The Bakers came, and would have carried all off at that Price, but the Amazonians swore, that they would carry the first man who attempted it before the Mayor; upon which the Farmers swore they would bring no more to Market; and the sanguine Females threatened the Farmers, that, if they did not, they would come and take it by Force out of their Ricks. The Farmers submitted and sold it for 6s on which the poor Weavers and Woolcombers were content.[26]

Cases from the early nineteenth century indicate the role of women in the 'rough music' of anti-poor law agitation, and in seasonal customary ritual such as Easter heaving or lifting.[27]

We can ask to what extent the recruitment of women workers in cottage manufacture and the putting-out arrangements of such industries relied on these pre-existing women's communities. How, otherwise, was a new cottage industry such as straw plaiting established, or an older occupation like woollen spinning introduced to a new area?[28] There is evidence, in the case of woollen and worsted spinning at least, that local women not only produced the yarn but acted as intermediaries themselves, 'putting-out' to other neighbours. For women were frequently convicted of embezzlement offences in the woollen and worsted industries. Most of the convictions were for bad spinning or false reeling;[29] women were convicted for embezzling themselves, and for failing to check the yarn they put out to neighbours.[30] We need to enquire further into the methods and avenues for the recruitment of women workers in both spinning and other cottage manufactures.

Plebeian community versus the market?

The role played by the impact of social relations established outside of the immediate workplace on the structure of work itself might suggest the existence of two spheres, one of moral imperatives and collaboration and another of commodity relations and economic individualism. We thus assume a dichotomy between the household or community and the social relations and imperatives of work. Olivia Harris points out that the categories of kinship and economics are usually regarded as mutually exclusive in terms of their moral axioms, and it is also clear that 'living in a single vicinity or neighbourhood has its own morality, with specific obligations and expectations, regardless of possible kin ties'. But the imperatives of the world of community cannot really be so separated off from the world of work, for both exchange and economic relations were significant in both worlds. As Harris warns,

Both the language of kinship and the way co-residence is represented, contain underlying assumptions about the exclusion of economic relations based on direct exchange and precise calculation, and the presence of other relations of generosity without calculation. This ideology . . . should not, however, be confused . . . with what relations actually obtain between kin and non-kin.

Kin and community convey to us a behaviour code of mutuality, but 'the degree to which people exhibit such behaviour to each other is a matter for investigation rather than assumption'.[31]

Economics was at the basis of mutuality just as much as of the market. In early twentieth-century London, neighbourhood sharing was viewed as exchange – 'in theory at least, reciprocity was the rule', and gifts created obligations.[32] And in eighteenth-century England the community consumption which Hans Medick has attached to E. P. Thompson's 'plebeian culture' was economic both in its manifestation and in its motivation. Time and especially money spent by the poor on cultural ritual, gifts, feasts and luxurious consumer display were a form of 'social exchange'.

> The social exchange, which was so typical an expression of plebeian culture, strengthened the bonds of kinship, of neighbourhood and friendship. Thus it produced just that solidarity to which the small producers could, in times of dearth, crisis and need, most easily have recourse.[33]

Consumption and exchange, activities usually considered in terms of economic categories, were thus characteristic features of the social relations of mutuality. But consumption was not just a narrow economic activity. It was a form of social participation, and as such it both governed and responded to the community networks we have discussed.

Anthropologists such as Mary Douglas treat consumer goods as an information system or means of communication. 'A household's expenditures on other people give an idea of whether it is isolated or well involved.' Consumption, and particularly luxury consumption, convey the fine gradations of social class, age and hierarchy as well as cementing particular kinds and degrees of social relationship. Goods are 'the medium, less objects of desire than threads of a veil that disguises the social relations under it'.[34]

Women and the organization of consumption

It can be argued that this consumer culture was very important to the community networks formed among women in the eighteenth century. The household production unit was also a unit of consumption. The consumer needs of the household had to be maintained and organized. To what extent was it the women who organized the household consumption, and indulged in the private and social luxury consumption? We know this was the case in early twentieth-century Europe. In London, 'wives' skills and tastes could do as much as husbands' wages to determine how comfortably their families lived'. Among the Amsterdam seamstresses the first task of a married woman was housekeeping. She saved on the family budget by sewing clothes for her family, and tidiness became 'the most valued quality in housekeeping'.[35]

In eighteenth-century England too, there is at least indirect evidence to show that women organized a large proportion of household consumption. It is evident that many of the new consumer industries reproduced goods which women already made for household consumption. Women's hands were busied producing yarn, stockings and clothing for their families. They also took pride in their labour-intensive efforts to by-pass the market and so to clothe their families better and with a smaller outlay of precious cash earnings. As noted in chapter 4, Eden found that whereas in the South most labourers purchased their clothing, in the North, 'Almost every family has its web of linen cloth annually, and one of woollen also.'

> It is generally acknowledged, that articles of clothing can be purchased in the shops at a lower price, than those who make them at home can afford to sell them for; but that, in the weaving, those manufactured by private families are very superior both in warmth and durability.[36]

In both England and Scotland most linen was made by private families for their own use. Though needlework and cooking had existed throughout the early modern period as the essential elements of housewifery performed by women, the demands made by these on women's time became more intensive and more highly skilled as new lighter materials in grades of cotton and linen, and new furnishings and cooking implements, were introduced over the course of the eighteenth century. The amount and the variety of household consumption increased, and women's household tasks increased with this. By the early nineteenth century long-outmoded industrial pursuits had been incorporated into labour-intensive housewifery carried to the extremes portrayed by George Eliot in her picture of the small tradesman and farming class in *The Mill on the Floss*. Bessie Tulliver exclaimed,

> To think of these cloths as I spun myself . . . and Job Haxey wove 'em and brought the piece home on his back. . . . And the pattern as I chose myself and bleached so beautiful, and I marked 'em so as nobody ever saw such marking. . . .

And this was linen produced not even for immediate use but as a kind of family treasure to be passed down to the eldest son.[37] The housewifery that became associated with the proliferation of home-produced and purchased commodities also became an indicator of a family's status. A woman's labour power was an important asset, but her consuming power for the household was also an asset of rising significance in the eighteenth century.

It is no mere coincidence that many of the new domestic manufactures of the seventeenth and eighteenth centuries were also consumer industries catering to a mass market, and that their labour force was made up predominantly of women. Joan Thirsk has argued that these are the industries which male historians have largely neglected.

> Starch, needles, pins, cooking pots, kettles, frying pans, lace, soap, vinegar, stockings do not appear on their shopping lists, but they regularly appear on mine . . . iron, glass, brass, lead and coal were important industries . . . but we are not yet sure that they employed as much labour or contributed as much to GNP in the seventeenth century as the common domestic

goods that were liable to turn up in every household of the land.[38]

At least one of these male historians has, however, attempted to make a case that the so-called 'home market' of the eighteenth century was largely a women's market. The consumer industries of the early Industrial Revolution were 'those in which women took the decision to consume: the cotton, woollen, linen and silk industries, the pottery industry, the cutlery industry, the Birmingham small trades'.[39]

We can ask to what extent women's organization of this household consumption created a consumer culture centred on the marketplace. Household management was also dependent on a knowledge of the price, a knowledge acquired through long-term participation in the market, and through the information acquired in the networks formed among consumers, particularly women. It was this process of haggling and bargaining of the market, according to 'that sort of rough equality which though not exact, is sufficient for carrying on the business of common life' which Adam Smith argued actually determined the extent to which the value of commodities accorded with their price.[40]

This consumer culture did not contradict the household economies and production of proto-industrial workers. It was but one part of household management. Similarly these proto-industrial workers and consumers were not obviously aware of any special distinction between the market and the moral economy. Many of the seasonal activities, rituals and customs were important sources of income in themselves, making it worthwhile leaving off waged work for one or more days at a time. Where waged work and household management inter-twined, other time and money economies took their own priorities, and these frequently concerned female members of the workforce.

One of those areas where capitalism and custom walked hand in hand was in luxury expenditure. Hans Medick argues that such expenditure was actually a communal manifestation which developed 'in harmony with the growth of capitalist markets'.[41]

172 *Manufacture and the Economy*

This luxury expenditure was for public display as much as for private indulgence. Eighteenth-century moralists singled out girls and women in their attacks on luxury expenditure, complaining of girls buying silk ribbons, hats, jewellery and dresses to suit every change in fashion. But men equally participated in the luxury consumption of dress. Skilled calico printers in the 'traditional, remote and parochial society' of Bury in Lancashire were 'a high and fashionable body of men' who 'displayed themselves at festivals in breeches, white silk stockings, silver buckles and powdered hair'. And Birmingham 'toys' were for male as much as female ornamentation – buckles, brooches, ornamental buttons, seals, snuff boxes and chains.[42] Spending limited money incomes on luxury conveyed social emulation and participation. The eighteenth-century social system was, as Adam Smith observed, a plutocracy, as much for the poor as for the rich. This spending was, moreover, the 'sporadic enjoyment of luxuries' in an uncertain economy of irregular income and underemployment.[43]

Luxury consumption and work discipline

This luxury expenditure was also a manifestation of the 'indiscipline' of labour in the eighteenth century. For workers 'wasted' time as well as money. Saving time and money was certainly important to merchants and manufacturers in the eighteenth century – the legions of their complaints make this clear. Time for them was capital; their profits were determined by the velocity of circulating capital, that is, the movement of stocks of goods tied up between stages of the production process and marketing. Yet workers and artisans in these early industrial communities appear to have failed to acquire a sense of the importance of this timekeeping. Saving time might have saved capital, but it did not contribute to the workers' subsistence or security. The so-called 'leisure preference' of early industrial workers was, according to Adam Smith, entirely rational. 'Our ancestors were idle for want of a sufficient encouragement to industry. It is better, says the proverb, to play for nothing than to work for nothing.' The casual work rhythms of the period, furthermore, found their context in the rhythms of home and family life, community and artisan society. For home and local community formed the location and framework of labour. Home and community set their own structure on work time, and the constraints of fairs, markets, raw material deliveries and putting-out networks imposed another strict discipline on labour.[44] The 'disciplining' of labour in the eighteenth century had its roots not in the factory system, but in the rise of new consumption patterns, and the organization of consumption as well as much production in the household. It was a discipline generic to the housewife and mother who planned day by day, week by week and far into the future.

Community relations and networks were integrated into the priorities of workplace relations not just, as in earlier periods, because the household was both the unit of production and the unit of residence. In the proto-industrial economy of the eighteenth century, consumption was the activity which bound community and capitalism together. The new industries produced consumer goods; they transformed goods formerly produced (largely by women) within the household to meet basic needs into commodities to be sold on a world market. The new industries also tapped a women's labour force, a labour force which brought valuable skills and social networks. It was also a cheap labour force, bound as it was within the household. But the communities into which this capitalist production penetrated themselves became consumer and market communities. Social status and participation, custom and community continued to hold sway and to impinge upon work, but they did so in new ways, in ways expressed increasingly through consumption, by the individual, by the household and by the whole community. The new industries which produced household consumer goods fed on a new household consumption organized by a labour-intensive housewifery.

Capitalist competition and market-orientated consumption certainly came to affect the structures and close ties between workplace and community. Population growth and migration could be effective solvents of corporate identity. So many regions of domestic industry had, by the late eighteenth century, been swelled by a dense, mobile and ceaselessly roving population. The ties among workers and among women were only one side of the story; divisions within these communities of small producers were just as important. There were the obvious divisions between artisans with a long and stable stake in the community or in a trade society, and casual outworkers in temporary residence. The bonds among women were important, but the divisions were also prominent. For female networks would be very finely tuned to the stage of the family life-cycle. There were the needs and priorities of those households of young women spinners who laid out the money they earned in domestic manufacture 'in buying fine clothes and other gawdy gew gaws'. And there were the conventional family priorities of older women such as those in the nineteenth-century Birmingham trades who took Saint Monday to do their laundry.

New directions

It seems, at first sight, that more long-term divisions created by the simultaneous process of population growth and increasing capitalist competition broke up the old bonds of custom and community. New production methods appeared divorced from community relations. It is said, for instance, that when calico printing came to Bury in Lancashire 'it was not long before the traditional revelries and entertainments of the village simply faded away before the approach of trade and unremitting employment'.[45] Similarly, in Birmingham, although the workshop remained resilient, the advent of steam power changed the rhythm of work, for though a worker could rent his own room and power he now had to work according to the hours in which the steam power was running.[46] But we have argued here that custom and community did not die away in the eighteenth century; rather they found a new context and manifestation through the market and consumption.

The impact of custom and community on the workplace was not, however, a casualty of industrialization; rather it took on other forms. Before we can understand the change which did come with the later phases of industrialization we must understand the content and dynamics of custom as well as the household itself. Appeals by historians to pre-industrial social values, nonmarket behaviour, family subsistence economy and backward-sloping labour-supply curves are all inadequate. Certainly the behaviour and characteristics subsumed under these terms affected the rhythms of work, the division of labour, and the use of and reception to new technology. But they were neither timeless not homogeneous and as yet we know very little indeed about them. One important aspect of these characteristics during the phase of proto-industrialization was the special integration of waged work, household subsistence and consumption, and community networks. It was women who filled the interstices of all these centres of activity. And it was the mixed character of women's household, waged and community work whose purpose above all others was to ensure the subsistence of their families, which made women workers so vulnerable to exploitation, and their labour such a lucrative source of profit to capitalists in the early stages of industrialization.

PART TWO

Paths to the Industrial Revolution

The regional and cyclical patterns of industrial advance and decline in the seventeenth and eighteenth centuries continued into the latter part of the eighteenth and into the nineteenth century. These brought with them new technologies which had far-reaching ramifications for the division of labour, including the sexual division of labour, and for community structures. But towards the end of the eighteenth century a rapid process of technological change in selected key industries cemented a new regional and industrial dominance. The paths taken by this new and unprecedented technological advance were as diverse as their industrial settings. The rapid development of a technology of handtools and small-scale machinery, the rapid proliferation of new hand techniques and skills, were just as notable as the more commonly recognized 'new technology' of mechanized steam-powered processes. Why these technologies arose as they did in their various industries is a question historians have long sought to answer. But looking at the question as they generally do, from hindsight, they have tended to assume the inevitability of but one striking path of technological change. The possibility of alternative paths which were, for some reason, blocked off in some industries, but allowed to develop in others, has rarely been explored. The way in which technology was developed within various organizational structures and devised to fit them – artisan or factory, home or workshop, centralized or decentralized decision-making – is rarely considered, for these structures are usually deemed to have been determined by the technology.

We must therefore examine the types of technological

advance which corresponded with specific structures of
industrial and work organization; we must show how these
structures adapted and developed their own technologies.
Posing a wide variety of economic structures, and with them
technologies, however, only tells us half the story. We must also
enquire into the mutability of these structures themselves in
response to the concomitant development of labour's skills and
forms of organization, and into the familial, cultural and
customary constraints on the workplace. These factors have
never been considered together: no wonder, for it would be a
mammoth task. But because they have only been considered
separately we have no real understanding of the relationship
between the economic and social history of work and the
process of technological change. We will, therefore, attempt to
examine some aspects of this relationship by looking at the
parallel but diverse development of two of the Industrial
Revolution's most important industries – textiles, and the metal
industries.

In the first instance we will look at present historical analyses
of technological diffusion. A survey of the various theories
appealed to by economic historians will be followed by a
consideration of the alternative Marxist analysis of technological
change and the labour process. The substantial contribution of
these theories to our understanding of the 'why' of technological
change still, however, begs many questions. The chapter will
raise those unanswered questions of how the social and
economic organization of industry – its industrial structure, the
gender, skills and customs of its workforce, and wider social
institutions – affected conditions of work and technological
change.

CHAPTER 8

The Economic History of Technological Diffusion

Technology lies at the centre of the Industrial Revolution, yet we still have little analytical understanding of the 'how, why and wherefore' of invention and diffusion. For technology is the economist's and the historian's 'black box'. It was fundamental to most other aspects of industrialization, but the contents of that 'black box' are difficult to unravel, and their structure is difficult to perceive. Traditional approaches and some more recent ones accepted the 'black box'; that is, they wrote of technology as an autonomous force. One such account, and an exemplary one, is David Landes's *The Unbound Prometheus*.

Landes provides a lucid and systematic account of the autonomous development of technology in the Industrial Revolution. His is a compelling picture of innovations proceeding forward, within and outside various sectors, in a logical challenge-and-response sequence dictated by the pressures and constraints of the 'interrelatedness' of techniques. He sums up the Industrial Revolution under three technological principles:

(1) the substitution of machines – rapid, regular, precise and tireless – for human skill and effort;
(2) the substitution of inanimate for animate sources of power – especially ways of converting heat into work;
(3) the use of new and more abundant raw materials. The substitution of mineral for vegetable or animal substances.

He shows how closely each set of technical changes was interrelated with the other. The breakthrough in textile machinery put pressures on raw materials, power sources and the engineering sector, leading in turn to a changeover from a wood to a coal fuel economy, to the introduction of steam as an immensely superior source of power and boundless source of energy, and to advances in engineering which brought standardization, uniformity and precision into machine making.

While possessing the power of simplification, this perspective also reinforced the old technological determination of most accounts of the Industrial Revolution. It was a determinism which was also peculiarly parochial in its account of the industrialization process. While presenting a survey of technological development in Europe, it followed the route of the most 'progressive' industries without enquiring into the reasons for the patterns of technological change across Europe's regions. While setting out the remarkable and seemingly unproblematic stages of mechanization in Britain, Landes does not ask why this mechanization took place so much more slowly in Britain than in the United States. In demonstrating the 'backwardness' and eventual upsurge of French industry, he does not ask why the new types of power and coal-using techniques so characteristic of the new British technology were so difficult to adapt to French production processes. Finally, he never asks what impact the new technologies had on skills, employment and labour conditions.[1]

The question of Britain's slow development was extensively debated among economic historians who challenged the view that technological development was autonomous. They looked instead to the economic reasons for why innovation took place in some countries, but not in others. The major economic factors behind the different technologies to be found across countries were the supplies of capital and labour. Economic historians now speak of the capital or labour 'intensity' of various techniques to describe a technology's relative use of the one or the other factor of production. And they argue that different relative endowments of, or costs of each factor of production between countries or regions either stimulated or retarded innovation. This more relativistic approach to the 'why' of technological change also meant more relativistic consideration of the 'effects' of the new techniques, because the new tools, machinery and processes frequently failed for some time to raise total product-

ivity significantly. This, in itself, was a legitimate reason for slow
innovation.

In turning to such debate, economic historians have made
much more explicit use of economic theory. But the old
economic history contributed much to the economic theory of
technical change. The traditional explanations have been
simplified, modified and recreated into new theories of technical
change. No longer trusting to the old traditions, however,
economic historians now turn to economic theory to give 'scientific' credibility to arguments which are not significantly
different from the old. But if the economic theory of technical
change has not offered a great deal that is new in explanations, it
has separated out and clearly specified some of them. Economic
historians now draw a distinction between those factors affecting
the invention of techniques from those affecting diffusion or
innovation. They concede that the springs of invention lie
beyond the limits of economic theory, but believe diffusion to lie
firmly within its grasp. Diffusion, in turn, is separated off from
the impact of technical change. The effects of new technology
are conceived to be microeconomic, that is, to affect costs,
prices, labour and total productivity in the single firm or industry; or they are conceived to be macroeconomic, to affect
growth, economic structure and employment. The jargon of the
economist has entered the language of the history of technology,
such that the diffusion of any technology is categorized into
'inducement mechanisms', 'induced bias', 'learning by doing',
'embodied' and 'intra-sectoral' innovation, and its impact
summed up in terms of the 'residual' input to economic growth.[2]
There may be some use to such exercises, but much more
interesting, it seems to me, were the areas where history and
theory interacted. One such case was the debate on the 'capital
or labour intensity' of new technologies.

Habakkuk and his critics

H. J. Habakkuk's classic formulation of the problem asked why
the British economy appeared to be so prodigal in its use of

labour, in contrast to the more mechanized state of the American economy.[3] The 'labour intensity' of British industry had long before been observed by Marx:

> Nowhere do we find a more shameful squandering of human labour power for the most despicable purposes than in England, the land of machinery.[4]

Habakkuk analysed the speed and character of technical change in both countries in terms, however, of differences not in the supply of labour but in the supply of land. The types of technologies developed in America economized, he argued, on labour because the American industrial wage had to equal average earnings in agriculture. The supply of labour was furthermore 'inelastic', that is, relatively unresponsive to small changes in the wage rate. Geography created great inflexibility in the labour market. In the words of one of Habakkuk's main critics, the essence of Habakkuk's thesis was this: 'If one country has a higher ratio of land to labour than another, all other things being equal, then this country will use more or perhaps better machinery for each worker in manufacturing than the other country.'[5] In Britain, with its small endowments of land, there was a large underemployed agricultural labour force with nowhere else to seek a better living except in industry. The general thickening of population in the countryside therefore made some areas suitable repositories for labour-intensive domestic industries.

Habakkuk did not confine his explanation of technological change to relative endowments of land and labour. For, as he recognized, all techniques depended on the use of some labour. Though some technologies required less labour overall, they may have needed more highly skilled labour. Habakkuk argued that the more capital-intensive techniques in America did indeed make use of more highly skilled labour, but this type of labour was also relatively cheaper in America than it was in Britain.[6] Habakkuk's argument made for a powerful model of the economic reasons behind the peculiar national characteristics of technological development, and of the reasons for the faster or slower rates of diffusion of particular techniques.

There was much, of course, which the model left out. First,

Habakkuk gave little weight to the impact of natural resource endowments apart from land, such as the effect of the opening up of the Ruhr on German industrialization. Second, when he wrote of the cost of labour he referred only to wage rates, failing to account for the costs of education and of the longer hours and faster pace of work. Third, mechanization might just as well take place because of cheap supplies of labour. For example, the introduction of the sewing machine coincided with the arrival in Massachusetts of cheap Irish labour, and it was this cheap labour that made a factory-dominated clothing industry possible.[7] The point might also be made that the adoption of any capital-intensive technique at one stage might just as well generate the introduction of labour-intensive processes at earlier or later stages of the production process. Some new capital-intensive techniques created demands for craft or domestic work where these had not existed before. And the close connections between the various stages of production meant that attempts to save labour or skilled labour at one stage through mechanization might involve the use of more labour or skills at earlier stages. The choice between techniques according to relative costs of factors of production was furthermore constrained by the structure of the market; for different techniques, particularly in the early phases of industrialization, often made for different qualities of product – higher or lower counts of yarn, coarse or fine cloth, more or less resilient crucibles, more standardized or highly differentiated commodities.[8]

Points like these only made special empirical qualifications to the hypothesis. Some neoclassical economists objected to the whole theory, for it did not fit the idea that innovation was determined by the relative prices of capital and labour. Habakkuk instead explained new technology by a complicated connection between the abundance of land, the high price of labour and the low cost of capital.[9]

The thesis was important to economists for the new departure it made from the neoclassical theory of technical change.[10] For it challenged the view that new techniques appeared in a continuous stream, and that they were the result only of entrepreneurs' attempts to reduce their labour costs. Habakkuk had seen the flaw in this view, and had argued quite correctly that the entrepreneur was more interested in reducing total costs than simply labour costs.[11]

While the theoretical questions raised by Habakkuk's thesis spawned a whole academic industry, the impact of this work was also to generate a series of case studies by American economic historians on the cost structures of various technologies with different rates of diffusion in Britain and America. The most challenging and best known of these studies[12] did not tell us anything about the origins of British technical change, for all were centred on the latter half of the nineteenth century, but the explanations they offered certainly could provide a basis for rethinking our traditional ideas about the origins of British technology. The purpose of these studies was to provide a critique of standard views on the theme of entrepreneurial failure in the late Victorian economy. Their perspective created a new and wider economic context to explain the so-called retardation of British technology at the end of the nineteenth century. Yet, with the notable exceptions of G. N. von Tunzelman's study of steam power and G. K. Hyde's study of the iron industry, a more critical view of the original speed of Britain's technological takeoff is yet to emerge.[13]

Factors which explain the slow diffusion of technology in the later nineteenth century can also help to explain earlier paths of technical change. We can, for example, compare cost and productivity schedules between old and new technologies, and so try to identify the threshold point where it was profitable to introduce new techniques. Such thresholds might be very high, for the costs of change were not only new machines but also new forms of work organization which went with the machines. The spinning mule, for example, drew on a different labour force than did the water frame or the jenny. The water frame, on the other hand, was introduced in large centralized 'factories', unlike the smaller workshops of the early jennies and mules. Later, powered weaving machinery entailed weaving sheds and other machinery such as dressing frames to prepare the yarn. These costs had to be added to the basic costs of the new machinery. A much later, but well-documented example, is that of agricultural machinery. The introduction of new harvest tools such as the scythe instead of the sickle, and ultimately the reaping machine, required a complete reorganization of the harvest.[14]

New techniques in many industries, furthermore, for some time required different types of raw material and generated different qualities of output. The jenny produced a fine weft, but no warp; the water frame produced a good warp, but inferior weft. The yarn which went to the power loom had to be 'dressed' first. Iron foundries used cheaper materials than the old wood-burning forges; but they produced pig iron which then had to be refined. If different techniques meant some, even slight, differences in output, market structures were to a large extent responsible for the use of newer, older or even compromise 'intermediate' technologies. Different counts of yarn, different qualities of cloth, standardized or quality design, all depended on different technologies.[15]

Social institutions are just as important to the reasons for different but coexisting technologies, as are the usual economic factors. But they are frequently difficult to specify. In nineteenth-century Britain entrenched historical institutions of property holding and enclosure, along with the layout of the fields, for a long time inhibited the use of reaping machinery. Different institutions of property holding between regions in the eighteenth century almost certainly affected industrial organization and growth; such institutions probably affected a region's reception to new technology as well.[16]

Apart from sociolegal institutions, there were also the institutions of employer-worker relations. Technologies which were associated with highly skilled male labour were also frequently constrained by strong trade unions or informal work groups. Employers in several cases sought to escape such constraints by using other techniques which drew on unorganized or unskilled female labour. Calico-printing technologies in the eighteenth century were developed against this backdrop, so too were the cloth-finishing techniques in the woollen and knitting industries which fired the Luddite episodes in the early nineteenth century. Later in that century, ringspinning was introduced in America instead of mulespinning for precisely these reasons.[17] Employers in textiles and engineering in this country clearly thought they could dispose of skilled labour by introducing a self-acting mule and semi-automatic machine tools.

Habakkuk's work and the many case studies of American technology which it inspired challenged old assumptions about the clear-cut differences between old and new techniques. The latest machinery was not always manifestly superior to the old tools, for the reductions in costs or increases in output to be had were frequently disappointing, or at least ambiguous. Real choices, in addition, had to be made over industrial organization, the division of labour and labour relations, and sociolegal institutions. British historians have not, on the whole, taken up the challenge to analyse these economic, social and institutional determinants of technical change. Two recent studies of the iron industry and steam power specify the strictly economic reasons for these innovations. Both argue, on the basis of a systematic analysis of quantitative data, that the cost of raw materials, in both cases that of coal, was the key consideration prompting innovation. Hyde argued that a sharp increase in the costs of charcoal smelting after the mid-eighteenth century coincided with a drop in the costs of coked pig iron. Very high profits went to those who decided on coke smelting, and in the 1760s average revenue exceeded average costs by £2 per ton. However, the speed of innovation was not just related to cost differentials, but also to prices, for the most rapid diffusion took place in the period of economic expansion between 1775 and 1815. Cost reductions and increases in demand both seemed exhausted after the Napoleonic Wars, though prices were still maintained by the expansion of foreign markets and the slow entry of new producers.[18] Von Tunzelman's similar study of the diffusion of the Watt steam engine presents the first detailed and systematic comparison of the fixed and variable costs of water power, the atmospheric and the Watt steam engines. He pointed to the long resilience of water power into the mid-nineteenth century, for wheels were cheap, lasted a long time, and called on little labour and no coal. The major costs were water rights and the high cost of installing the wheels. The so-called steam revolution of the 1840s and 1850s was accompanied by only a limited decline in the use of water power – a decline of 9 per cent in the British textile industry, and 14 per cent in Britain as a whole. There was not only a complicated choice to be made between steam and water power, but among different steam engines themselves. In spite of the historical fanfare over Watt's engine, the old atmospheric engine held its own. The speed of diffusion of the Watt engine over the atmospheric was determined not by technical superiority but by the higher fixed costs of the Watt engine, the length of time the patent had left to run, and the level of coal prices in the region. Most of the heavily industrialized areas of Britain were regions of cheap coal, and so for some time at least there was little real need to introduce the newest steam technology.[19]

These new and more critical studies of two of the most celebrated technologies which are virtually identified with the first Industrial Revolution convey some of the economic complexities which underlay the process of technological transformation. They convey a picture of a transformation which was gradual or cyclical, and not instantaneous or revolutionary. But the brunt of their argument still remains within the rather narrow framework of cost-accounting and price responses. Though a great deal can be gleaned about the path of innovation by use of an argument of basic economic rationality, this is an approach which assumes perfect knowledge on the part of its historical actors, and furthermore assumes perfect transferability of labour, capital and other inputs. Labour and capital are treated as abstract entities, and no mention is made of the manufacturing organization or of local customary and historical traditions which could either ensure or long delay an innovation's successful introduction.

Institutional factors

Quantities of capital and labour are only one side of the coin. There is also the quality of capital and labour; there is the institutional or customary context in which these 'factors of production' grew. Nathan Rosenberg, for one, pointed out the existence of different national technologies.[20] The character of a nation's technology could have important effects on its future capacity for invention and innovation. There were historical reasons, he thought, why nations with a more capital-intensive technology grew faster. Many of the human skills important to economic growth were acquired in the production and employment of a capital-intensive technology. The development of capital-intensive techniques became associated in time with a strategic inventive role for the sector producing capital goods. The development of such a sector which improved the efficiency of producing capital goods would, in time, become a source of capital-saving over the economy as a whole. A special sector producing capital goods was able to develop a base of technical skills and knowledge; it became the institutional means of transferring techniques to other sectors, and of developing new ones. Furthermore, a relatively small number of broadly similar production processes might be developed and spread over a large number of industries. The skills acquired in the production of capital goods and intermediate products could be used throughout the economy.[21]

A special industry and the institutions associated with producing capital goods were important, but even more significant in the early years of the Industrial Revolution were the skills and the labour which made these capital goods. Skills in processing and working metals, skills in engineering enabled a new technology to be effectively introduced and to work. Techniques which relied on coal fuels, like iron puddling and new ways of making glass, as well as those which used steam power, were dependent for their introduction, success and repair on new and adaptable skills as well as an old-fashioned practical 'knack' and know-how.[22]

The development of technology is not a self-fulfilling prophecy. At many points there are choices to be made, and these choices have been influenced by market structure, differences in final products and resource costs. Such choices were, however, constrained by institutions and customs which favoured certain kinds of innovation and skill. Economists and historians have, however, reached an impasse with this discussion. Highly abstract but very narrow economic analysis imposed a new economic determinism to replace the old technological determinism. Rosenberg and other economic historians recognized that social institutions were important, but they still kept to a narrow terrain. Only recently has debate moved to the whole sphere of industrial structure and workplace relations. The development of technology was shaped by these factors, as much as it was by conventional economic factors. But economic theory did not give historians the framework they needed in order to include these other more elusive and social factors. And they turned, instead, to Marx.

The labour process and the Marxist alternative

Instead of asking what narrow economic costs influenced the diffusion of new technology, a new generation of radicals demanded a total rethinking of technology in relation to manufacturing organization and to the character of work itself. While economists' awareness of the role of raw material costs, labour and capital costs, and demand factors challenged the old technological determinism, it blinkered consideration of the historical contexts of technical change in the pressures and constraints of the wider economy and society. Economists ignored the actual process of technical change within the workplace.

The new challenge was, therefore, to specify the connections between technical change and parallel changes in workshop and factory organization, to relate these to hierarchical organizations in production, and to analyse the impact of a skilled or unskilled workforce. The real questions at issue behind technical change and workplace organization concerned those who claimed the gains of change, and who controlled the pace and direction of work. Where economists regarded technology as an artifact, that is, a newer or older machine, radicals saw it as a process – a combination of tools, machinery, skills, work practices and organization through which production took place.

Two classic texts inspired a whole new enquiry into the source and effects of technical change – Stephen Marglin's 'What Do Bosses Do? The Origins and Functions of Hierarchy in Capitalist Production', and Harry Braverman's *Labor and Monopoly Capital*. The first challenged conventional ideas on the origins of the division of labour and the factory system, arguing both were introduced not for reasons of efficiency, but because they offered the capitalist the means for greater control of his workforce and an opportunity to claim a higher proportion of the surplus. Contrary to the accepted view that the rise of the factory was caused by the introduction of power-driven machinery, Marglin dismissed the so-called 'technological superiority' of the factory and with this its technological origins. Factories existed well before powered machinery, and what was really at stake in the Industrial Revolution was not efficiency, but social power, hierarchy and the discipline of labour. Marglin also pointed out, and has recently emphasized, the way the factory itself became an impetus to technological innovation. For capitalists sought out and developed techniques which were compatible with large-scale factory organization. The water frame was an example. Originally designed as a small machine turned by hand and capable of being used in the home, it was patented by Arkwright and only henceforth built as a large-scale piece of machinery driven by water or steam power. Marglin thus argued that even though the factory did not actually determine prevailing forms of work organization, capitalist control and machines were nevertheless most highly developed in the factory form of organization. It was this high degree of capitalist control in the factory which in turn constrained the development of technology.[23]

Harry Braverman's *Labor and Monopoly Capital* also formulated connections between changes in technology and work organization, but he looked at phases of mechanization as an aspect of the history of the rise of scientific management. Taking Paul Baran's thesis of the rise of monopoly capital, he described the growth of the modern corporation in terms of the rise of automation. His book went back to the Industrial Revolution to seek the origins of scientific management and Fordism, in order ultimately to comment on the implications of a new computer revolution.

Both these texts were published at a time when advanced computing technology and a new microelectronic revolution of the 1970s and 1980s were first becoming apparent. Implications for employment, job structures and manufacturing organizations were predicted to be unprecedented. It was this economic and social context of our own industrial age, as much as the new perspectives of Marglin and Braverman, that challenged radicals to take account in their political analysis of the enormous changes taking place at the basic level of the workplace. Marxist historians thus turned their attention to the 'labour process'. What, then, is the Marxist theory of the 'labour process' which underlay the radical analysis of technological change?

Marx defined the labour process as the basic relation between man and nature which exists in all modes of production. The labour process has three constituents: the work or labour itself, the subject of labour or nature's materials which are worked up into raw materials, and the instruments of labour which mediate between labour and the subject of labour. In the capitalist mode of production, the elements of the labour process are combined to produce surplus value as well as use values, so that the value of the commodities produced in the labour process is greater than its constituent elements. The basic dynamic of capitalism is founded on the drive to increase this surplus value. Marx divided the ways of increasing surplus value into two categories. The first was those ways of increasing what he called absolute surplus value. This was the formal subordination of labour power, that is, labour was coerced into producing more with the same techniques through means to increase the speed and intensity of labour, to introduce longer hours, or to expand the scale of production. The second was ways of increasing relative surplus value. This was the real subordination of labour power, and was carried out through gains in productivity, the introduction of machinery, and the conscious application of science and technology.

Marx connected these ways of increasing surplus value to historical phases in the development of capitalism. These were:

first, cooperation, where a number of labourers working together produced more than when each worked separately; second, manufacture, distinguished by increases in productivity through division of labour; and finally, modern industry, identified with the introduction of machinery to replace labour power and to raise the productivity of the workers who remained. Manufacture described the process through which the division of labour splits up productive activity into component parts, separating workers into skilled and unskilled, thereby creating a hierarchy of labour powers. But in manufacture workers still had some degree of control over the content, speed, intensity and rhythm of their work. In modern industry, the capitalist took this control. For with the use of machinery, work activities could be homogenized, and workers deskilled. Workers became mere appendages of the machine, and capital could appropriate to itself all the functions of specification, organization and control. The historical phases of capitalist production also coincided with the use of different means of increasing surplus value. Cooperation and manufacture were dominated by means for raising absolute surplus value, and these were ultimately limited by the length of the working day. The possibilities for raising relative surplus value which were most obvious in modern machine industry were, however, limitless, for they proceeded through increases in productivity.[24]

This Marxist framework was an important alternative to conventional economic theory, and it inspired a rethinking of technological change among radicals and historians. Most of the first applications of Marx's theory of the labour process, and, with this, responses to and criticisms of the work of Marglin and Braverman, were focused on American examples of technical change and workers' struggles.[25] But a number of recent studies of later nineteenth-century changes in technology and work organization have applied the idea of the labour process, or have made central to their analysis the struggle between workers and employers over the exercise of control at the workplace.

Marx's discussion of the labour process opened a new historical and social dimension in the study of technological change, but the 'labour process' contained many problems of theory and historical validity. Many Marxist historians, whatever may have been the intentions of Marx himself, gave overriding significance to the effects of machinery, or, in Marxist terms, to the 'forces of production'. First, the accepted division between absolute and relative surplus value hinged almost entirely on the introduction of machinery. But Marx's own discussion of the labour process stresses the equally important place of the 'intensification of labour', that is, harder and faster work.[26]

With this in mind, we may cast the empirical validity of 'modern industry' in a new light. What was striking in nineteenth-century Britain was not the power of the machine in displacing labour, but the dependence of most work processes on more and more labour. Even after the Industrial Revolution, the labour process remained 'dependent on the strength, skill, quickness and sureness of touch of the individual worker, rather than on the simultaneous and repetitive operations of the machine'. And where machinery was introduced it 'created a whole new world of labour-intensive jobs'.[27] The reasons for this were not, however, just the conventional economic factors of labour supply, profitability and cost of production, market uncertainties and preferences, technical difficulties and the availability of some intermediate technology. They were also the successful use of the 'intensification of labour' to raise surplus value, and the workplace and wider political disputes over machinery, the length of the working day, and the division and speed of labour.

These disputes produced some remarkable detours and by-passes in the routes of technological change. Wide political struggles in the first half of the nineteenth century to reduce the length of the working day culminated in the factory legislation of 1847 which cut the hours of work in factories from anything up to sixteen hours down to ten. This sharp inroad on conventional means of raising absolute surplus value resulted, however, in the first rapid innovation of the high-pressure steam engine. If the workday was to be cut, manufacturers responded in droves by seeking the means of increasing both the intensity and the productivity of the working hours which remained. Study of the numbers of new and adapted engines introduced, the greater speed of machinery, and the increase in the number of accidents through boiler explosions has established this new burst of innovation.[28] Was its timing, soon after the ten-hours legislation, just coincidence?

Such workplace and political struggle could mean incentives to the development of new and other technologies, as they did in this phase of the steam revolution.[29] But, equally, they could limit the implementation of changes in the structure of the labour force, or the organization of the workplace which new technologies might allow. The self-acting mule was one such technology. Employers in the 1820s and 1830s actively sought this invention to rid themselves of troublesome skilled mulespinners. Andrew Ure virtually regarded the self-actor as the salvation of capitalism, and Marx several decades later accepted that the self-actor had robbed a particularly militant group of workers of jobs and skills. But, in fact, the self-acting mule did not displace the skilled mulespinner, for existing workplace organization embodied in the minder-piecer system remained and successfully blocked a redivision of labour which might have allowed capitalists to use workers with lower subsistence wages. Yet simultaneously employers found other means of increasing their surplus value through the practice of 'time cribbing', that is, out-of-hours cleaning and oiling of the machinery. Further technical improvements in the self-actor allowed it to be built longer and run faster, and many firms practised 'stretch out' and 'speed up'.[30]

Workplace organization and strategic struggles by workers to deal with the threats to jobs and skills brought by mechanization mattered in many industries. Skilled workers succeeded in maintaining control in the printing industry, in spite of new technology; but they failed to do so in engineering.[31] These are only several examples, among many, of the very different outcomes of workers' struggles.

Marxist dilemmas

Marxists using the labour-process analysis have, however, been overly impressed with the machine, failing to understand the place of the speed and division of labour. In some industries, it was the latter which created the crucial changes that revolutionized an industry. The building industry is a good example. Historians always write about the traditionalism of the building trades, for little new machinery was introduced in the last century. But the key to changes in the production process was not machinery, but the rise of general contracting from the 1830s, that is to say, a transition in the organization of work.[32] Marxists have also accorded far greater power and success to machinery than workplace struggles actually allowed. The old linear framework of technological determinism challenged by economists has not, in effect, been entirely expunged by the Marxists. For though they argue that technological change is the outcome of struggles between workers and capitalists, their search for examples of deskilling, divisions of labour and mechanization in any historical period is inspired by questions and interpretations of production and work suitable only to modern Western capitalist economies. They furthermore seek to situate their individual studies in terms of key turning points that marked out the transition to manufacture or modern industry as the case may be. The result has been a failure to grasp the diversity of the experience of industrialization. There were many alternatives to mechanization in improved hand technology, the use of cheap labour-saving materials, and the division of labour and simplification of individual tasks, which were developed in their own right. The factory, furthermore, was but one form of work organization among several, including the putting-out system, small workshops and artisan production.[33] But what matters, surely, is not the fact of the coexistence of hand and machine techniques, and centralized and decentralized processes, but the range between these polarities and the different uses and therefore meanings of each. Our task is to look at their relations to each other and the kinds of social and economic pressures which stimulated the cyclical phases of development and the regional concentrations of these many diverse forms of technology and work organization.

We must seek to find a nonlinear perspective, one which shows not just the relations between technology and work organization, but also the range of capitalist forms of development. The existence of hand technologies and artisan work organization also reveals the existence of alternative forms which could be exploited just as effectively as the factory and the machine to raise capitalist profit and control. The extent to which this was achieved, however, was a question of struggle within each technology and form of work organization between employers and workers' organizations, customs and institutions.

Another problem of the labour-process perspective is its narrow focus on the workplace and production process. The impact of culture, community and family in the workplace itself is ignored. This is a peculiarly 'male' perspective, and it is not surprising that virtually all of our historical studies of labour processes are focused on a male labour force and male attitudes to work. Although recently social historians have stressed the role of the world outside of work in shaping the structure and attitudes of the so-called labour aristocracy,[34] their example has not been followed in studies of work and technical change. Most of these studies have focused on male industries or only the male workers in those industries – iron and steel, printing and engineering, mulespinning and building. For such industries exemplified the struggles of the skilled craftworker against the inroads of machinery or deskilling. Most have failed to realize that in many industries the term 'deskilling' meant the introduction of women workers. The attributes of the skilled craftworker were still acquired through apprenticeship, 'independence' (high enough wages to support himself without recourse to charity or poor law), mobility, mutuality, collectivity and the unspoken but assumed virtue of masculinity.[35] These ideals were all well and good, but from the viewpoint of only one part of the workforce. What did these ideals look like from the perspective of the ethnic minorities, the women, the unskilled poor who were excluded from the jobs, the public houses and the social institutions inhabited by the skilled man?[36] Analyses based on the division of labour must take account, not only of changes in men's jobs, but of implications for the structure of family employment and, in most of the textile and metalworking industries, of divisions between men's and women's work.

This account of the histories and theories of technical change highlights the recent questioning of the neutrality and the 'inevitability' of technological change. Technology, the classical example of a 'black box', is now being opened and its contents scrutinized by economist and social historian alike. But the bearing on technology and work organization of a wider framework of social institutions and customs, the different meanings and values attached to work and production in different historical settings, have as yet received but little attention.

CHAPTER 9

The Textile Industries: Organizing Work

The experience of growth, decline and transformation of the eighteenth-century textile industries frequently passes for the whole story of Britain's economic revolution. Of course, this means a very one-sided, and indeed blinkered, view of economic fortunes and prospects in the period. But we cannot deny the significance of textiles to the British industrial experience. Behind their significance, too, lay a richly textured saga of growth and decline, small-scale industry and large-, home and factory, manual labour and machinery. For the textile industries include far more than the cotton manufacture: there was the experience of work in wool and worsted, linen, silk and framework knitting. How were these other branches of the textile industry organized in the eighteenth century? What were their tools and techniques, and what happened to them? We will here compare the progress and decline of several branches of the textile industry in the eighteenth century – not just the success stories, but the failures and those which 'also ran'.

Though most historians have been mesmerized by the remarkable growth of the cotton industry at the end of the eighteenth century, what is striking to the historian who takes a longer and broader view over the century is the substantial and impressive growth of all the major textile industries – wool and worsted, stocking knitting, silk and linen as well as cotton, and with this the rapid complementary spurt of calico printing with the rise of cotton. We have already compared output and productivity across the main textile industries; now we shall look at the origins and forms of development of the various forms of work organization.

What is clear at the outset is the great variety of pre-industrial and proto-industrial structures. The story of textiles, the epitome of the whole story from proto-industry to Industrial Revolution, is frequently caricatured as a series of transitions from the artisan to the putting-out system, and thence to the factory. But in fact, features of all these types of work organization and various permutations of them existed from the very beginning of the eighteenth century within and between the various textile industries. Proto-industry did not, in effect, take on any single type of organization; nor for that matter did it entail any special type of technology.

Industrial origins

Wool and worsted

Before cotton we generally think of wool. What were the main centres of this traditional industry in the eighteenth century? For most of the time, the woollen and worsted industries were spread widely over the country, but they were also specialized by region. Defoe in 1726 found broad cloth and druggets in Wiltshire, Gloucestershire and Worcestershire, serges in Devon and Somerset, narrow cloth in Yorkshire and Staffordshire, kerseys, half-thicks, plains and coarser things in Lancashire and Westmoreland, shalloons in Northampton, Berkshire, Oxfordshire, Southampton and York, worsteds in Norfolk, lindsey woolseys at Kidderminster, flannels in Salisbury and Wales, and tammeys in Coventry.[1] By 1770, the wool-manufacturing centres stretched from Exeter through Witney and Leicester, out to Newtown, up through Bradford and Kendal and from Galashiels to Aberdeen.[2] In 1792 the inhabitants of most areas of the North and Scotland were occupied in some part of the wool or worsted industries, and there were more so employed in Leicestershire, Oxfordshire, Derbyshire, Norwich and many areas of the West Country.[3] By the last third of the century the real centres of the woollen and worsted industry had concentrated in East Anglia, the West Country, the East Midlands, Yorkshire, Lancashire, the north Pennines, and the Scottish borders, but Yorkshire's predominance was already noticeable. Between 1741 and 1772 the input of raw material into the industry had increased by 14 per cent per decade, and by 1770 the output of the industry was valued at £8–10 million. Yorkshire then accounted for one third of this value and for one half of the value of all textile exports. In effect the main expansion of the whole industry in the last third of the century was accounted for by the rise of Yorkshire and Lancashire, and it was from the 1770s that the great divide between the West Country and Yorkshire opened up. Between 1770 and 1800 the proportion of wool textile exports going to America rose from 25 per cent to 40 per cent, and virtually all of this increase was supplied by Yorkshire.[4]

Yorkshire's rising supremacy was based partly on wool and partly on the newer worsted manufacture. The manufacture of worsteds had spread first in the sixteenth century with the introduction of the New Draperies in Norfolk and spread further in the seventeenth and eighteenth centuries with the manufacture of bays, serges and shalloons. It spread rapidly in the West Riding in the eighteenth century; Halifax was the key centre until succeeded by Bradford during the Industrial Revolution.[5]

Yorkshire's rise meant Norfolk's demise, but this prospect was by no means apparent for much of the eighteenth century. Certainly Norfolk's worsteds were of a high quality, but fashions eventually changed against the heavily glazed materials in favour of finer merino fabrics with silk decorations.[6] And by 1770 Yorkshire was producing worsteds to a value equal to Norwich's.[7] Norfolk's original success was, however, based on its successful competition with the West Country woollen industry, for it produced goods at a total cost of some 8 to 10 per cent less than the West Country's, and its weavers' wages in 1760 were 40 per cent lower. Its woollen and worsted industries were expanding rapidly in the first half of the eighteenth century, and at its height Norwich commanded 12,000 looms and 72,000 weavers working to the order of thirty large cloth dyers. The industry grew until the 1770s, after which it went through periods of strength and activity until its collapse in the early nineteenth century.[8]

Colchester, Suffolk, Coventry, Worcester, Dorset and Exeter

had likewise all been flourishing centres of the cloth industry in 1700, but by 1800 all were far in decline. The course of decline in Essex was fairly typical. The cloth industry there in 1700 dominated four large towns and a dozen small towns and villages. It contributed to the employment of the majority of Essex families, for most women in the towns and the countryside were spinners. But after 1700 rural weavers were going into rapid decline, and the smaller centres became the first casualties. Local capital was gradually shifted away from textiles and into farming.[9]

The well-established pre-industrial cloth industry of the West Country remained buoyant throughout most of the eighteenth century, but by its end the trade was split. In the area around Stroud in Gloucestershire there was very small expansion, while in Trowbridge and Bradford-on-Avon on the Somerset-Wiltshire border a transfer of production into cassimeres, a fine twisted cloth, cancelled the decline in broad cloth.[10] Trowbridge in fact prospered, its population rising by 57 per cent between 1811 and 1821. It was the most prosperous centre of the West Country industry, until it too declined in the later 1820s.[11]

Yorkshire's apparently meteoric rise in the eighteenth century was founded on a long period of apprenticeship reaching back to the fifteenth century. Halifax wares were sold at the St Bartholomew's Fair and at the Blackwell Hall market in London. By the seventeenth century Wakefield and Leeds were the great wool and cloth markets of the area. Some part of the woollen manufacture was carried on over the whole of the North, the West and parts of the East Riding, though it was very diffused compared to the Leeds, Halifax and Wakefield regions. The main worsted area stretched from Bradford to fifteen miles west and northwest of Halifax, taking in the upper valleys of the Aire and Calder. Halifax, Keighley, Haworth and Colne were centres, with a considerable worsted manufacture also found in the area of Leeds and Wakefield. The woollen district stretched in a pentagon between Wakefield, Huddersfield, Halifax, Bradford and Leeds, with the great cloth market at Leeds.[12] The industry was largely organized in villages which in the seventeenth and eighteenth centuries had grown at the expense of the towns, and even by 1811 most towns were small and contained only one quarter of the population of the whole West Riding.[13]

Knitting

The woollen and worsted industries were complemented by another widespread domestic industry in the seventeenth and eighteenth centuries – the hand- and subsequently machine-knit stocking industry. Joan Thirsk has described the rise of the fashion for silk, wool, worsted and cotton knitted stockings, and the wide geographical dispersal of the handknitting industry by the end of the seventeenth century. Although in many areas knitting complemented the woollen industry, the more significant common factor was a large population of smallholders pursuing pastoral farming.[14] Handknitting in Wensleydale and Swaledale was an extension of the old Westmoreland industry, and by the eighteenth century desperation pursued the villagers.[15]

In Richmond (Yorkshire) 'every family was employed great and small' in the manufacture of knitted yarn stockings for ordinary people. Doncaster's handknitting industry was famous even in the sixteenth and seventeenth centuries, and it was an industry almost entirely in the hands of women.[16]

Machine knitting came to the East Midlands in the mid-seventeenth century, where it first diffused as the occupation of yeoman knitters of some substance in the villages of south Nottinghamshire, Derbyshire and Leicestershire. But the industry was rapidly urbanized, particularly after the mass migration of framework knitting from London to the Midlands in the first half of the eighteenth century. There were less than 12 frames in Leicester in the 1680s, but by 1700 the town had 600. The village industry also continued to grow – 16 per cent of the inhabitants of Wigston Magna between 1698 and 1701 were knitters, while the proportion of knitters in Shepshed rose from 4 per cent in 1701–9 to 25 per cent in 1719–30.[17]

Silk

Machinery came early to the knitting industry, but the development of technology and organization in the silk industry was even more precocious. After the introduction of Lombe's silk-throwing machinery in 1719, mills sprang up all over the country; by the late eighteenth century the industry was still scattered over twenty counties and fifty towns. The largest eighteenth-century silk mill was in Stockport with six engines and 2,000 workpeople. Together with other smaller mills it supplied the Spitalfield weavers.[18] Tightly capitalist from the outset, silk throwing spawned important weaving and ribbon-weaving sectors in London and Coventry. The Coventry ribbon-weaving industry, in particular, combined traditional artisan structures with new capitalistic methods. In the eighteenth century, the industry moved into a town with long-established textile traditions in the manufacture first of blue thread, then of woollen and broad cloth. But it also spread in periods of prosperity among the wives of the colliers in villages to the north and northeast of the city, and this was soon an area where 13,000 looms supported 30,000 people.[19]

Linen

The rapid growth and early capitalistic structures in both silk and in framework knitting were matched by another but equally impressive combination of growth and family production in the linen industry. The eighteenth century saw a rapid increase in the colonial demand for linen for slaves' clothing, coffee and indigo sacks, and mattress covers, as well as in the domestic demand for the 'decencies' made from linen – tablecloths and napkins, towelling, bedding, furnishings, and clothing, especially shirts.[20] An import rather than an export industry, linen accounted for 15 per cent of total imports in 1700, but only 5 per cent in 1800, and sources of supply shifted from the continent to Ireland, Scotland and domestic linen production itself.[21] It was an industry with a long history of small-scale production for localized markets. In one sample taken from probate inventories of the late sixteenth and early seventeenth centuries, 14 per cent of agricultural labourers were engaged part-time in working up flax, and a further 15 per cent in working up hemp. The industry flowered between 1740 and 1790 under trade protection. A good deal of it was still hidden away, incorporated into domestic production for family use:

'Tis true the English manufacturer is not publically known, or at least not so much taken notice of as the Scotch or the Irish, but the reason of this is very plain: in this country most of the linen we make is made by private families for their own use, or made and consumed in our country towns and villages . . .[22]

The commercial industry also provided for several regional economies. It was the staple industry in parts of Yorkshire and County Durham, in several parts of Lancashire ranging from Lancaster and Preston down to Manchester, and in the non-woollen area of Somerset and Dorset, spreading into Devon and Wiltshire.[23]

The Scottish and Irish linen industries arose as staple industries to supply the substantial English demand after English protection sharply reduced continental imports. In Scotland, the industry was encouraged by the new trade opportunities created by the Act of Union, with help from the Board of Manufacturers which brought over several cambric weavers from France and a man skilled in all branches of linen from Ireland to travel about and instruct weavers in the trade. A visitor to the Highlands in 1725 wrote that 'Every woman made her web and bleached it herself and the price never rose above 2s a yard, and with this cloth almost everyone was clothed.' The manufacture was well established in Aberdeen and the countryside by 1745; by 1795, 10,000 women were spinning yarn, and 2,000 women and 600 men were employed in the thread manufacture in Aberdeen. It was a staple industry of Glasgow from 1725, and other major centres were found in Forfar, Fife, Perth and Dundee.[24] Even Edinburgh had a high-quality branch. The Irish industry similarly rose to prominence on the basis of English demand. Between 1740 and 1770 cloth exported rose from 6.6 million yards to 20.6, and yarn exports rose from 18.5 to 33.4 thousand cwt. Seven-eighths of these exports went to Britain. The export market in yarn was further stimulated by the expansion of the English cotton industry, which for the first three quarters of the eighteenth century depended on linen warp for the manufacture of calicoes.[25]

The linen industry also fostered the growth of the cotton industry. The two industries lived together along with a hybrid, the fustian manufacture, for a long time in Lancashire, and in other areas such as Glasgow the cotton industry drew on the older linen industry's skilled labour force. Duties against foreign linen as well as the early eighteenth-century prohibition of Indian printed calicoes stimulated both British manufactures.

> The cotton industry was enabled to grow dramatically once its inherent technical potentialities were realized because of its long roots in the linen industry. In Lancashire (as in Lanarkshire) there were established skills to be drawn upon and there was a nationwide commercial superstructure of provincial drapers and London warehouse men.[26]

Cotton

As significant as this complementarity between linen and cotton was the developing connection between calico printing and cotton, for a large part of the origin of great demand for cotton textiles in the eighteenth century can be explained by the enormous popularity of the fashion for printed fabrics which emerged in the later seventeenth. Calico printing, originally established in Eygpt, was soon relocated to the main importing centres for London – Amsterdam and Marseilles. It was carried across Western Europe by the Huguenots and was soon also found throughout Eastern Europe. And the British industry, except for the luxury end of the market, soon migrated from the metropolis to the provinces. Chapman and Chassagne have recently demonstrated the significance of the stimulus provided by this industry. By 1792 nearly a million pieces of white cotton cloth were produced in Britain, of which 60 per cent were sent to the printer's. Calico-printing workshops were effectively 'protofactories', a 'transitional stage in evolution from dispersed domestic manufacture to the factory system'; and a number of leading calico printers were associated with the introduction of mechanized spinning and weaving.[27]

Although calico printing was prohibited in Britain in 1720, printing on linen or cotton-linen mixture was a popular alternative, and the cotton industry soon grew out of the linen or linen mixtures industry. In mid-eighteenth-century Lancashire, a master 'put out' linen yarn for the warp and cotton weft in cops. The spinning was arranged by the merchant or carried out by the weaver's family, and cotton spinning formed the part- time activity of most women of the labouring classes. In Scotland, the weavers of Paisley and Glasgow, already skilled in the production of fine linens, easily turned their hands to fine cottons, notably the Paisley shawls. In the 1770s several thousand looms in the Glasgow area were producing linens, silks, cambrics and lawns; they changed over by the end of the century to fine cottons.[28] Glasgow concentrated on plain and printed muslins and Paisley on fancy fabrics. In Lancashire spinning spread early in the south around Manchester, while weaving was done on handlooms in the northeast corner of the county.[29] The impact of the production of calicoes and muslins was described in 1785 in MacPherson's *Annals of Commerce*.

A handsome cotton gown was not attainable by women in humble circumstances, and thence the cottons were mixed with linen yarns to reduce their price. But now cotton yarn is cheaper than linen yarn, and cotton goods are very much used in place of cambrics, lawns and other expensive fabrics of flax; and they have almost totally superseded the silks. Women of all ranks, from the highest to the lowest, are clothed in British manufactures of cotton . . . the ingenuity of the calico printers has kept pace with the ingenuity of the weavers and others concerned in the preceding stages of the manufacture, and produced patterns of printed goods which, for elegance of drawing, far exceed anything that ever was imported; and for durability of colour, for generally they stand the washing as well as to appear fresh and new every time they are washed, and give an air of neatness and cleanliness to the wearer beyond the elegance of silk in the first freshness of its transistory lustre . . .[30]

By the end of the eighteenth century the geographical distribution of most of the main textile industries had undergone great change. Yorkshire now dominated the woollen and worsted industries, though Norfolk and the West Country remained in a strong yet static position. Knitting was now centred on the East Midlands, particularly Leicestershire and Nottinghamshire after framework knitting had shifted out of London, and the formerly widespread country handknitting industry was now more localized in the North and in Scotland. Silk remained a small luxury industry, though highly organized along highly capitalistic lines, and received a new fillip in the eighteenth century with the expansion of weaving in London, and later ribbon weaving in Coventry. Linen and cotton became highly concentrated – the one in Scotland and Ireland; the other in Lancashire and Scotland. All the branches of the textile industry in fact went through significant phases of expansion in the eighteenth century, but the remarkable spurt of the cotton industry was unique and soon cast the respectable performances of its predecessors into shadow.

Early work organization in textile manufacture

The organization of production in the early stages of the different textile industries shaped their subsequent paths into industrialization. Elucidating the variety of industrial structures prompts us to seek out the reasons for difference. Mercantile capital intervened in all these industries, but with very different implications for organization. There was in some industries a straightforward correlation between capitalist control, frequently in the form of concentrated ownership, and putting-out systems. But other industries, while using mercantile networks, were run by independent small clothiers who preserved artisan structures. What contributed to the fragility or resilience of these proto-industrial structures? Why did capitalist control intervene more effectively in some, but not other of these textile industries? The answer to these questions may lie to some extent in costs of production or market structures. But they are likely to lie at least as much in social structure and institutions.

Wool and worsted

At the beginning of our period, the wool and worsted industries in both the West Country and Yorkshire contained an abundance of small clothiers. Julia Mann has argued that until the last half of the eighteenth-century the small clothiers of the West Country were a significant part of the local social structure. There was no rigid dividing line between these and other workers; they went poaching in company with cordwainers, shearmen, bakers and glaziers; shopkeepers often carried on a little clothmaking.[31] Heaton describes the textile class of seventeenth-century Yorkshire as mainly small clothiers making one piece of cloth a week and living hand to mouth. There were also yeomen who combined agriculture and industry, either making or finishing cloth, and then there were the large clothiers whose chief interest was cloth manufacture. They were mainly found in the villages near Leeds, where they also kept gardens, orchards and closes for animals. They kept a full set of clothmaking utensils and employed journeymen, women and apprentices. The larger clothiers often bought pieces from small men and sold these along with the cloth of their own manufacture to merchants from London and Yorkshire. The cloth was then taken once or twice a week to the open markets in Leeds, Halifax or Wakefield, or sent in cargoes to Blackwell Hall or Bartholomew Fair. The small clothier's establishment was largely determined by the size of the labour force needed to make a piece of cloth; in the case of the kersey manufacture it took six people sorting, carding, spinning, weaving and shearing for one week to produce one finished but undyed piece.[32]

During the course of the eighteenth century a division arose in Yorkshire between the woollen and worsted branches. The small independent wool clothiers remained much as they had been in the seventeenth century. The father went to market and bought the wool; the wife and children carded and spun it, and some of the wool was put out to be spun in neighbouring cottages. With the help of his sons, apprentices or journeymen, the clothier then dyed the wool, wove it, took it to the fulling mill, and then to his stall in the market. He produced only one or two pieces a week, and kept between three and fifteen acres of land. Some had a horse or ass to carry their burden to market; others carried the pieces on their head or shoulders. It still only cost between £100 and £150 to start out, and the system of open marketing placed the small producer on equal terms with the large.[33]

But this division between wool and worsted was regional before it was industrial. In the more populous areas of the West Riding, especially in the area near to Halifax which was soon to turn to the worsted manufacture, Defoe found the cloth industry organized in a highly sophisticated combination of domestic and workshop manufacture. He found the country 'one continued village' with hardly a house standing out of speaking distance from another; 'at almost every house there was a tenter and almost on every tenter a piece of cloth, or kersey or shalloon'. For two to three miles in almost every direction 'look which way we would, high to the tops, and low to the bottoms, it was all the same; innumerable houses and tenters, and a white piece upon every tenter'.

> Among the manufacturer's houses are likewise scattered an infinite number of cottages or small dwellings, in which dwell the workmen which are employed, the women and children of whom are always busy carding, spinning etc. . . . this is the reason also why we saw so few people without doors, but if we knocked at the door of any of the master manufacturers, we presently saw a houseful of lusty fellows, some at the dye-fat, some dressing the cloths, some in the loom, some one thing, some another, all hard at work.[34]

The worsted manufacture, based on combed rather than carded long-fibred wool, was organized in Yorkshire on a much more capitalistic basis right from the outset. There the small independent clothier never existed; instead there were merchant-manufacturers who resembled the large West of England clothiers. They bought large quantities of wool at big fairs and put it out over a wide area to be spun and woven. The open marketing at the Leeds cloth hall with over 1,000 stallholders was in sharp contrast to the concentrated industrial structure indicated by the worsted hall at Bradford with its 250 stallholders. The Yorkshire worsted manufacturers ran extensive putting-out networks, commonly distributing wool within a radius of twenty to thirty miles. Woolpacks were often consigned to shopkeepers or small farmers who received a sum for delivering and receiving wool and spun hanks of yarn. 'The mother or head of the family then plucked the tops into pieces the length of the wool, and gave it to the different branches of the family to spin about nine or ten hanks a day.'[35]

Capitalistic structures were, however, by no means necessary features of the worsted manufacture, for in Norfolk independent craftsmen were the order of the day in the industry. In Norwich the weaver was the pivot of the industrial structure. He bought yarn from independent spinners and wove it by himself or with his journeymen who worked either on commission or on the weaver's premises. The cloth was then put out to independent cloth finishers, and afterwards sold locally to drapers or sent to London. Neither, for that matter, were artisan structures natural to the woollen industry. The West Country by mid-century was the living example to contemporaries of monopoly and capitalist putting-out systems.[36]

The sharp contrast between the artisan and capitalist structures of wool and worsted in Yorkshire, of worsted in Norfolk and Yorkshire, and wool in Yorkshire and the West Country cannot be attributed to natural differences in the industries. Neither do such differences account for the early appearance of mills and factories in the different textile manufactures. Mills existed in both industries from the early eighteenth century. And artisan and capitalist alike both made use of water-powered factories, or at least mills. Such mills were used for specific processes of manufacture and were generally incorporated into existing artisan and putting-out structures. Water-powered fulling mills existed from early times, but these were not regarded as factories.[37]

Do labour costs help to explain the different structures of wool and worsted? This seems unlikely, for labour costs were said to be lower in Norfolk than in the West Country, and they were lower in turn in Yorkshire's worsted manufacture. Putting-out systems prevailed in both the West Country and the west Yorkshire worsted manufacture in spite of different labour costs. Markets, to be sure, favoured the new over the old: first worsteds in Norfolk, then those in Yorkshire; and Yorkshire's wool over the West Country's. Guild, corporate or landed regulations played their part in all three regions, but with different results. Putting-out systems in West Country wool and Yorkshire worsted probably owed much of their origin to concentrated ownership; and the local origins of such concentration go back in turn to a range of structural and circumstantial social factors peculiar to each region.

Framework knitting
Artisan and capitalist organizations grew and developed together in the woollen and worsted industries, though both industries were carried out in different regions, even within Yorkshire itself. In framework knitting capitalist relations were an historical development of the eighteenth century. Framework knitting started out as a skilled occupation practised by yeomen of some substance, similar in character to the peasant metal-workers around Sheffield. Early frames were cheap, and in the villages workshops of four to six frames were often built as annexes to houses, with larger production units in the towns. The frame was still, however, costly compared to the capital stock of a handloom weaver. It ranged in price from £3 10s to £18, though many used secondhand frames. The earlier machines of the seventeenth century had cost a great deal more and taken two men to run. Master knitters operated an apprentice-journeyman system, and their capital outlay compared to that of a cutler in the eighteenth century. The technology was soon simplified, and many were able to build their own frames. Many rural stockingers until the later eighteenth century worked three or four days a week at knitting, and carried on another occupation such as farming. During the first phase of its migration to the Midlands the industry was located in villages of middling wealth and relatively egalitarian social structure. Many villages which took up framework

knitting had previous traditions, not so much in handknitting as in woollen or worsted weaving; others were located in the vicinity of a metalworking area, as was Nottingham, where stocking machinery was first developed.[38] Knitting villages soon became subject to the pressures facing many areas of subdivided tenancies where a large class of smallholders was emerging.

In the eighteenth century framework knitting became very closely connected with developments in both the silk manufacture and the cotton industry. For it was the stocking frame which created the possibility of abandoning clumsy woollen hose in favour of lighter and more elegant stockings of silk and cotton. The main centre of the industry, originally in London and controlled by the London Chartered Framework Knitters' Company, quickly spread to the Midlands in the later seventeenth and early eighteenth centuries; by the middle of the eighteenth it was focused on Leicester and Nottingham. The wealth and power of the Midland hosiers expanded rapidly as they responded to the caprice of eighteenth-century changes in fashion, using new materials and creating new meshes and new garments on the frame. The London Company's attempts to maintain apprenticeship regulations in the trade throughout the country were ignored by the Nottingham hosiers. Many of these had not themselves served a legal apprenticeship, and they employed unlawful journeymen and women and children in large numbers.[39] The knitters complained of masters who could 'build fine houses and country villas, keep carriages and equipages, go a hunting etc., while begrudging a trivial rise to their workers'.

Apart from sophisticated putting-out networks, the industry also boasted centralized workshops from early in the eighteenth century. Large workshops employing over forty parish apprentices existed in Nottingham from the early 1720s, and Samuel Fellows, one leading hosier, had built a large 'safe box' factory to manufacture imitation Spanish lace gloves in 1763.

Thus by the time that Hargreaves and Arkwright went to Nottingham, the concentration of both juvenile and adult labour in factories was a fairly familiar idea. The way was prepared for further development of factory industry.[40]

Concentration also extended to entry into the industry. Working framework knitters rarely found their way into the ranks of the hosiers. These latter formed an elite which moved into the same residential districts as the gentry and took their recruits from among the sons of 'gentlemen, farmers and prosperous tradesmen'.[41]

In framework knitting, as in wool and worsted, neither labour costs nor even capital costs contributed seriously to concentration. But the circumstances of markets, and the high incentive to by-pass the old guild controls did open opportunities for a limited number. Reinforced in turn by regional poverty founded on agrarian change and population growth, the hold of the larger clothiers was soon assured.

Silk

Silk was a luxury industry, but it spawned the country's first factories – highly capitalistic enterprises employing child labour – and one of the country's most highly skilled and traditionally organized artisan groups, the Spitalfields weavers. Capitalist and artisan confronted each other across the throwing and weaving branches of the trade. And even artisans eventually divided, separating themselves off from degraded outworkers in a split between metropolis and province, and between town and country. The throwing mills which sprang up around the country after Lombe's Derby mill first appeared, employed young girls and children, but worked to supply a traditionally organized and skilled craft weaving branch. The industry grew up against the backdrop of a series of acts to prohibit the importation of silk goods. In Spitalfields, London, the industry was organized on an artisan basis. Most of the householders were small master weavers who sold their output to mercers or drapers, who in turn retailed these to private customers in City shops. A seven-year apprenticeship prevailed and masters kept two to three journeymen by the year. Spitalfields was known at this early stage for its manufacture of elaborate brocades, damasks, velvets and other rich fabrics. Most of the other silk-weaving centres which became well known by the end of the eighteenth century grew up in the wake of the Spitalfields Acts of 1773.

Under the provisions of these acts the wages of silk weavers were to be fixed in London by the Lord Mayor, recorder and aldermen, and in Middlesex and Westminster by the magistrates. In 1792 the acts were extended to silk mixtures and in 1811 came to cover women as well as men.[42] The acts were established after a period of falling wages and violence in the trade. Samuel Sholl wrote of the early 1770s:

> But in process of time, as there was no established price for labour in England, there was great oppression, confusion and disorder. Many base and ill designing masters took the advantage, in a dead time of trade, to reduce the price of labour. The oppression became so insupportable that a number of journeymen, at the hazard of their lives, resolved to make examples of some of the most oppressive of the manufacturers by destroying their works in the looms. This they effected, but for want of prudence in their conduct, several fell victim to the cause and lost their lives.

The effect of the acts, however, was to prompt many manufacturers to move their trade to other districts. The silk-button trade moved to Macclesfield, a town whose throwing mills were already providing the Spitalfields trade. Silk weaving spread to the villages nearest the East End of London in the late eighteenth century, and by the early nineteenth into the villages of Essex. Silk ribbon weaving became important in Coventry from the beginning of the nineteenth century, but was built there on the basis of an older silk manufacture going back to the seventeenth. As the worsted manufacture moved from Norfolk and Suffolk, and as the conditions of the Manchester cotton weavers deteriorated in the early nineteenth century, silk weaving moved in to take its place.[43]

The structure of the Coventry ribbon-weaving industry was perhaps one of the most interesting developments to emerge from this regional diversification. While stocking knitting started as a skilled yeoman's trade which was soon to face a rapid decline in status, ribbon weaving, which developed later, faced pressures which were quite different in the town and country branches of the industry. The trade was important only from the early nineteenth century and was concentrated in Coventry and a number of outlying villages in a twelve-mile radius, including Nuneaton, Foleshill and Bedworth. A distinction between the towns and the villages quickly arose. The firsthand journeymen in the Hillfields area of Coventry were well off, while the trade in the villages was poor, degraded and carried out largely by colliers' and farmers' wives and children. The villagers had no piecelist and were banned from the use of more efficient looms. Weavers in the city and its suburbs were considered to have 'superior habits and intelligence' to the dispersed and ignorant inhabitants of the rural parishes. These were employed chiefly in the singlehanded trade and retained most of 'their original barbarism with an accession of vice'.[44] The strength of the urban weavers was largely established through the concentration of the industry in the hands of a small number of manufacturers, and through the weavers' success in preventing an influx of cheap labour. The industry was dominated by only a dozen families whose control extended from the earliest days of silk production to the late Victorian period.[45] The master manufacturers were able to control the industry at least until 1812 by means of the undertaking system. The manufacturer provided the silk dyed in the hank to the undertaker who provided looms and either did the work with his family or was assisted by apprentice and journey hands. Male and female journey hands were required to serve a five- to seven-year apprenticeship.[46]

Linen and cotton
The linen industry was first and foremost a home occupation, widely practised, even after its commercialization, as a basic part of household duties. Never defined as a skilled occupation, it was largely assumed to be a women's preserve. Much of the domestic English linen manufacture was submerged in private family production and use. In Scotland, too, 'Many of the Scotch ladies are good housewives, and many gentlemen

of good estate are not ashamed to wear the clothes of their wives and servants' spinning.'[47]

While putting-out systems went hand in hand with industrial concentration in the other textile branches examined thus far, the cotton manufacture tells a different story. Here, from its earliest days a putting-out type of organization emerged in a more dispersed industrial structure. The manufacture intermediate between linen and cotton was fustian production, a mixed cotton and linen cloth. It was a minor cloth until the mid-eighteenth century, but the rural workers producing it were more caught up in capitalist relations than either linen workers or many wool producers. By the middle of the eighteenth century there was systematic intervention by middlemen and a developed putting-out system, but large merchants did not control the markets or prices of yarn. Fustian masters appeared in the middle of the century. These gave out raw cotton and linen thread to the workers, then sold the cloth so produced to merchants. It seems probable that this prevalence of the small yeoman capitalist was an important factor in the successful growth of the Lancashire cotton trade. He could obtain credit or mortgage his land and become a putter-out. From here he could rise to become a small-scale employer of weavers and thence to the status of merchant.[48] These fustian masters or factors were generally responsible for a large body of small weavers over a wide geographical area. The system had advantages, for the factors did most of the 'managing' of the industry, leaving merchants to concentrate on selling.[49] In spite of this putting-out structure, however, the organization of the industry was hindered by the large number of middlemen, the seasonal labour in agricultural districts, and the delays caused by the small scale of production.[50]

The rise of some of the really big cotton masters and their innovation of the early factory system was closely bound up with the profits of fashion to be gained in calico printing. When calico printing moved beyond London to Lancashire in the mid-eighteenth century, those who took up the business were drawn from the same stratum of society as the cotton spinners, flax spinners and merchants – 'the chapmen or dealers in linen or cotton cloth known as Blackburn Greys'. It was the middlemen suppliers to the Blackburn merchant houses – the Claytons, the Liveseys, Peels, Howarths and others – who started out in workshops which grew rapidly in size and efficiency. Peel had early connections with carding and spinning innovations through one of his spinners, Hargreaves, and pressured him into giving up the secret of his invention. Arkwright too was closely involved with calico printing. The increase in the demand for calico to produce the extremely popular printed cloth must have brought great pressure on the cotton spinners. In fact, as Chapman has shown, Peel devoted years to producing the finest possible fabrics for calico printing. He experimented with carding, roving and spinning machinery, building up his own team of artisans to build the machines. By the 1780s fine spinning was virtually an integral part of calico printing so that most manufacturers were calling themselves 'calico muslin manufacturers' or 'calico printers and muslin manufacturers'. And by the 1790s the spread of calico printing was putting a strain on the output of handloom weavers, so that it now became the general practice for manufacturers 'to establish their weaving . . . in all the little villages about [their works] in some of which they put a manager and take apprentices and also give out work to the inhabitants at their houses'.[51]

The calico-printing workshops, or 'protofactories', formed a nucleus for a series of other workshops gathering together hand technologies, and the cotton industry developed through a combination of dispersed and concentrated, factory and putting-out, forms of production employing a complementarity of mechanized and hand technologies. When Arkwright's water frame mills started to appear there already existed small but much more widespread jenny factories and small carding factories preparing cotton for home spinning.[52]

But in spite of factories, a substantial amount of spinning, both of high and low counts, was still carried out at home, or in very small factories. Some of these sold their yarn to the larger spinning factories, bridging any gaps during cyclical, technical or labour disruptions.[53]

Where industrial concentration, not production costs or markets, accounted for the predominance of putting-out networks in Yorkshire worsteds and framework knitting, this was not apparently the case in the early days of the cotton manufacture. Here, the market and pre-existing mercantile networks appear to have played the vital role in introducing a new product. Hence the production of this new material was diffused through the putting-out networks of the commercial linen and calico-printing industries. Capitalistic structures came with the new product; they were not imposed upon it. But the market opportunities created by a new product also entailed more open ownership and easier entry than did the concentrated putting-out systems of older textile sectors. This openness was also clearly related to regional social and institutional structures – the absence of corporate regulation over most of the cotton region, and the opportunities for population increase without the widespread poverty experienced in the East Midlands.

The impact of social structure
The proto-industrialization of the British textile industries in the eighteenth century was thus no singleminded progression, for the starting points of each industry spanned the whole spectrum of work organization. The classic early pattern, combining small independent artisans with agriculture and textiles, existed in some branches of the woollen industry, notably in Yorkshire, but not so much in the West Country; it existed in handknitting and some early framework knitting, and in linen. Some of the textile industries were, however, much more capitalistic, if not at the start then certainly by the middle of the eighteenth century. Putting-out and centralized processes prevailed from early on in the Yorkshire worsted manufacture, in the fustian manufacture, in framework knitting, silk throwing, calico printing and cotton.

Why were these structures so different even before the onset of severe pressures from technical change and capitalist competition at the end of the eighteenth century? The origins of these differences lie in several factors. The market played its part, and so did technology. But most influence was exercised by regional and industrial social structures, social structures which went back to the very character of feudalism and the origins of agrarian capitalism.

In the silk industry and in calico printing, a luxury market and traditional urban artisanal social structures encouraged a paradoxical combination of capitalistic organization and guild controls. Silk mills and calico-printing workshops were among the most advanced of early factories. But the strength of guild regulations helped to create metropolitan and provincial town and country divisions in both calico printing and silk weaving; divisions which were reflected in the quality of output and the labour force.

The putting-out system prevailed in the Yorkshire worsted industry, the West Country woollen industry, the Midlands framework-knitting industry, and the early cotton industry. The reason again lay largely in local social structures. This is clearest in the divide between the artisan woollen region and the putting-out worsted regions of Yorkshire.

Overpopulated pastoral regions marked by social division also help to account for the social structure of the framework-knitting industry. By the mid-eighteenth century the cost of frames combined with a relatively poor workforce to make for the easy concentration of frames in the hands of putting-out masters. This capitalist concentration and a large, flexible and weakened labour force created ideal conditions for the proliferation of capitalist structures of work organization. Internal control over the size of their labour forces prevented the growth of such capitalistic structures in industries such as the Yorkshire woollen sector and the Coventry silk-weaving trade. Coventry's weavers were protected by a ring of common land over which all master weavers had a right of pasture. This restricted the growth of the town and reinforced the strength of the urban weavers.[54]

What is remarkable is that textiles, the group of industries always singled out as the leader of the Industrial Revolution, and the prototype of advanced factory organization, should not only have started with multifarious forms of work organization, but should have retained these throughout the period of industrialization. Even by the 1820s only some stages of a few of these textile industries relied on factory organization and mechanized power technologies. The significance of these processes and industries, as in cotton spinning, should not, of course, go unrecognized. But it was still the case that decentralized, work-shop, artisan and putting-out systems were successful and profitable, and in addition a substantial degree of technological change was compatible with these structures.

New work organization and the road to the factory system

The artisan systems, putting-out networks and different types of early factories were all transformed over the course of the late eighteenth and the first years of the nineteenth centuries. The textile manufactures constituted the industry *par excellence* in that large-scale technical change that we call the Industrial Revolution. But it was competition and capitalist pressures, and not new technology itself, which accounted for the new forms of work organization evolving by the end of the eighteenth century. Some of the older types of work organization did develop into a factory system; others never did. Instead they developed their own valid and competitive forms or went into industrial decline. Technological development, on its own, did not have a great deal to do with the outcome. Many of the new techniques developed in the latter half of the eighteenth century could have been adopted across several systems of work organization. Yet some were adopted in only one of the forms they might have taken. This was the case with the development of the water frame and calico-printing techniques inside a factory system. We will discuss their cases more fully in the next chapter. But equally, the development of new technology in this period did not just mean mechanized, power-using techniques. It also included labour-intensive technologies and improvements in hand technologies which could be used in either domestic or factory production.

The particular technological developments of the various textile manufactures and the problems of their reception and diffusion merit a separate discussion, but here we will wish to see the historical development of the industrial structure of the textile industries. What happened to the artisan workshops, the putting-out systems and the cooperative factories by the time that the Industrial Revolution reached its peak?

Wool and worsted

The sharp contrast to be found in the early organizational structures of the Yorkshire woollen and worsted industries was intensified during industrialization. The putting-out system of the worsted industry developed into a factory system divided between large mill owners and poor wage labourers. The artisan system in the woollen industry was retained until the mid-nineteenth century, but adapted to needs for space and power by introducing cooperative mills for the use of all small clothiers who subscribed.

Earlier organization in the woollen and worsted industries in Yorkshire first appeared to be moving in quite different directions than this. By 1800, there were a large number of woollen mills containing fulling, scribbling and carding machinery, as well as a number of hand processes; whereas mills were only slowly developed in the worsted industry, partly because of labour opposition to the factory system, and partly because the industry was located in areas with little water power.[55] There were a number of large woollen clothiers around Leeds in the late eighteenth century who ran large workshops resembling miniature factories – James Walker of Wortley had twenty-one looms, eleven in his own loomshop and the rest in weavers' houses. One L. Atkinson of Huddersfield had seventeen looms in the one room. The reasons given by these particular clothiers for direct supervision were 'to have the work near at hand, to have it under our inspection every day, that we may see it spun to a proper length', and 'principally to prevent embezzlement, but if we meet with men we can depend on for honesty, we prefer having the cloths woven at their own houses'. From the 1790s, however, one hears many complaints from domestic clothiers, particularly in the worsted industry, that merchants were becoming manufacturers and so ruining the small independent men.[56]

The development of the factory system in woollens and worsteds ultimately, however, tended to much greater concentration and larger scale in worsted than in wool. The early mills in the woollen industry were not a challenge to but a part of the traditional artisan structure, and most of these mills as well as a number of later cloth mills were 'occupied and run if not entirely financed by small manufacturers previously involved in proto-industry rather than by wealthy mercantile owners'. Water-powered fulling paved the way for the centralization of other processes: soon the water wheels which turned the fulling stocks were turning new scribbling and carding machinery, and eventually even the water frames for spinning worsted.

Both systems of production – artisan and capitalist – could, therefore, generate centralized production processes, in effect forms of a factory system. But these were factory systems which clearly differed in their social relations of production. It was possible, on the one hand, from within and while preserving the artisan system, to make a gradual transition from cottage and cottage workshop to a factory with all the processes of production under one roof. But it was equally possible, on the other hand, for factories to be founded by wool clothiers with the stated purpose of greater supervision, quality control and prevention of embezzlement. Contemporaries were clearly aware of a big distinction between the two types of establishment. The former was generally referred to merely as a 'mill', that is, a centre where domestic manufacturers could bring their own materials to be taken through the mechanized processes of scribbling, carding, fulling and so forth. The latter was truly a factory – an establishment where a manufacturer employed wage-earning labour and with mechanical power made up his own raw materials into yarn and later into cloth. These differences were also reflected in the workforce. By the 1830s, 75 per cent of factory workers in the woollen manufacture were men with experience in the craft sector. Worsted mills, which had a much higher threshold of capitalization, were generally owned by putting-out merchants. The workforce was predominantly female and juvenile, and there was little upward mobility.[57]

The division within the woollen industry between Yorkshire and the West Country also grew wider during industrialization. Where earlier in the eighteenth century there had been a number of small clothiers in the West Country, these appear to have been forced out during this period. At the end of the century there were a number of attempts to exclude the small men coming in on the basis of factors' credit, and a new and very marked division arose between the 'respectable' and 'inferior' clothiers. By the early nineteenth century the small clothier had disappeared in Wiltshire and Somerset, and was replaced by large factories supplying domestic weavers.

Framework knitting
Whereas the putting-out networks in the Yorkshire worsted industry, and to a lesser extent in the West Country woollen industry, developed into a factory system, and in the West Country ultimately into de-industrialization; in framework knitting they developed into sweated industry. Technological change did to a certain extent increase capital requirements, but most of the increase in capital actually sprang from changing organization in the industry. From the mid-eighteenth century hosiers shifted from small workshops annexed to their homes into greater complexes including house, workshop, warehouse and rows of brick cottages built for their framework knitters. There was also a new class of bag hosiers renting out frames. One such hosier, Francis Beardsley of Bromscote, died in 1763 possessed of 112 frames in knitters' houses in Nottingham and 25 in villages in the vicinity.[58]

In spite of the earlier existence of open villages and substantial yeoman-artisan connections in south Nottinghamshire, Derbyshire and Leicestershire, framework knitting did not behave like the artisan-agricultural base in the West Riding woollen industry. Instead it declined into a degraded putting-out industry. The structure of the open villages in this area was perhaps not so restrictive as to prevent the influx of population which undermined the artisan. Leicester, for instance, had no belt of common land surrounding it, as had Coventry, to prevent migration from the countryside. And certainly from the early to mid-eighteenth century, a number of hosiers were employing an unrestricted number of apprentices as well as waged labour. Middlemen appeared in the industry from the mid-eighteenth century, leading to a reduction in knitters' income by 20 per cent, and technical change in the frame again in the mid-eighteenth century implied a further reduction in the stockinger's status. By this time the industry was spreading to poorer villages, so that, for instance, in Shepshed it was the poverty of the peasants in the second quarter of the century which formed a precondition for the area's subsequent indus-trialization.[59]

Silk

Work organization in the silk manufacture was closely tied to conditions in Spitalfields. The decline in the status of the Spitalfields silk weavers by the end of the eighteenth century was hastened in the early nineteenth by the spread of silk weaving about the country in a manner similar to the traditions of silk throwing. Throwsters in Macclesfield now produced for local weavers in the silk-button trade. Those in Leek provided for a family industry carried out in garrets. The industry in Manchester and Essex fed on cheap labour thrown off first by the woollen and worsted industry, and later by cotton handloom weaving. This provincial, as well as foreign, competition had reduced the Spitalfields weavers by the early nineteenth century to sweated labour, and the status of those in Macclesfield and Manchester was soon to take a similar course, after the rapid entry of cheap labour from cotton handloom weaving in the 1820s.[60]

Artisan traditions remained much stronger, however, in the Coventry silk-weaving industry. They were protected by the control exerted over population increase in the city, as well as over entry to the trade. Workers' resistance long prevented the introduction of steam power; no one even attempted to set up a steam factory until 1831, and this was immediately burnt down. Another built in 1837 survived, and after this a number of factories did appear in the city, but the suburb of Hillfields remained an artisan stronghold. The industry, however, became increasingly divided between town and country. Restrictive practices, highly skilled labour and a high-quality manufacture were the hallmarks of the town, while the villages in a twelve-mile radius around Coventry took the unregulated end of the trade, using largely women workers and less efficient techniques. But even the urban trade fell prey to capitalist pressures in the early nineteenth century. After the Napoleonic Wars a number of small masters appeared employing women and half-pay apprentices. And divisions also appeared among the journeymen. There were the firsthand journeymen who owned their own looms and owned or rented their houses. In 1838, 1,828 of these, 214 of them women, owned 3,967 looms. Then there were the journeymen's journeymen who worked either for firsthand journeymen or in factories. These accounted for 1,225–1,878 men and 347 women, and with their families made up 2,480 workers, of whom 373 worked in factories and the rest for the firsthand journeymen.[61]

The factory system was not the clear-cut outcome of industrialization in any of these textile sectors, except for Yorkshire worsteds. Instead, earlier industrial structures were simply intensified – responding to the pressures for capitalist expansion by accommodating these to artisan institutions, or by giving way to the full exploitation of sweated labour. But factories clearly did appear in the linen and cotton industries. Why did cotton in particular take this new direction, and just what kind of break with the old did it entail?

Linen and cotton
Putting-out systems and artisan production likewise prevailed in the linen industry for most of the eighteenth century, and even after the factory appeared at its end the hand production of linen goods continued to be a staple element of noncommercial or extremely local household production. Putting-out networks and factories came to the linen industry with the commercialization and localization of the industry, but the change was gradual.

The first linen mills were not built in Scotland until the

1780s; they remained insignificant until the first decade of the nineteenth century. Here, as in Ireland, after factory spinning started to displace women's domestic spinning, women were deflected into handweaving, and an oversupply of labour in weaving was already apparent by 1815.[62] Dundee, well known from early in the eighteenth century for coarse linens, was a case in point of one extreme of the variable development of industrial organization in the industry. Until the beginning of the nineteenth century yarn was typically spun by housewives in the countryside who then brought the yarn into Dundee to sell. But manufacturers, finding it difficult to acquire uniform sizes and qualities, started to use agents who purchased directly from householders. A few spinning mills did make their appearance at the beginning of the nineteenth century, but they were not significant until the 1820s. By 1822 there were seventeen steam-powered flax mills in Dundee, employing 2,000 people, plus another thirty-two mills in the neighbourhood.[63]

It was in the cotton industry that new organization and industrialization seemed most clearly united. But here too the break with older forms of work organization was not really so marked. As in the other textile manufactures, industrialization brought the intensification of a number of pre-existing forms of work. The domestic system easily accommodated the small mule and jenny factories which first appeared in the 1770s and 1780s. These small factories were set up in hamlets where they formed the nucleus of a village. An example was the village of Cheadle Hulme whose factory in 1777 contained two carding engines, five spinning jennies and one twisting jenny, though the building was large enough for fifteen jennies, rovers and carders.[64] The jenny mills ranged from the small shop with a hand carding machine to the water-powered factory containing all preparatory and finishing processes.[65] Spinning mules, at first hand- or horse-powered, were also operated early on in very small factories. Even in 1800 many mulespinners had set up in converted premises, and in a town like Oldham, which later became dominated by very large manufacturers, there were large numbers of small mulespinning entrepreneurs. These had raised their capital on the strength of connections with the land, with coalmining, or with the domestic textile manufacture.[66]

It was the Arkwright-type mills which marked the really big divide. These were thousand-spindle mills built to use Arkwright's newly patented water frame. But as will be argued more fully later, this centralization was not caused by the development in technology. It was caused by key business decisions. And these Arkwright mills were still located in the countryside close to sources of water power. They were closely interconnected with the small jenny and mule shops and with domestic weaving run on the putting-out system. Converted premises and mills with many tenancies were in fact the most dominant form taken by the factory system in the cotton industry in the eighteenth century. The first mills in Stockport were silk mills which subsequently declined, to be converted to cotton manufacture. The water power, buildings and child labour were all simply transferred from one textile industry to the other.[67] In addition to conversions there was the widespread practice of renting parts of mills.

Cotton weaving remained a workshop or domestic trade until the power loom started to spread more rapidly after the Napoleonic Wars. But industrialization made important inroads on the workplace of the handloom weaver long before the power loom became a real threat. There was already an important demarcation between the urban and rural branches of the trade. In the towns workshops were substantial, and a journeyman-apprenticeship system prevailed. The Manchester smallware and check weavers were already well organized in trade societies by the 1750s. They were weaver-artisans, self-employed and working by the piece for a choice of masters. The rural fustian weavers, however, were poverty-stricken outworkers.

This rural-urban division in the trade became blurred with the increase of population and influx of immigrants into south Lancashire in the latter part of the eighteenth century. We read a great deal of the declining status of the urban weaver, of his attempts to enforce apprenticeship restrictions and to establish a minimum wage. But we know a great deal less of work relations and the extent of the putting-out system among rural weavers.

We do know that the expansion of the cotton industry in the later eighteenth century attracted large numbers of new weavers from amongst small farmers, agricultural labourers and immigrants. 'It was the loom, not the cotton mill, which attracted immigrants in their thousands.'[68]

In the process, rural weavers became increasingly dependent on 'putters-out' who took yarn into the uplands, or upon particular spinning mills, for many early cotton manufacturers claimed to pay hundreds of handloom weavers scattered through the countryside in addition to the employees in their own mills. The dependency and outworker status of these weavers soon also affected the urban weavers. 'The artisan, or journeyman weaver, becomes merged in the generic hand-loom weaver . . . the older artisans . . . were placed on a par with the new immigrants.'[69] Poor outworkers were a cheap and flexible source of labour. Where concern over quality or timekeeping demanded something more, handloom weaving sheds, easily accommodated to existing mill sites, served the purpose.

Factories and alternatives

We have now unravelled some of the threads of development of the diverse structures of the textile industry over the course of the eighteenth and early nineteenth centuries. Let us therefore try to gather together these threads in order to capture the underlying pattern of the web of the eighteenth-century textile industries. The general effect of capitalist competition and technical change at the end of that century seems to have involved an intensification of existing differences in the manufacturing structures of the textile industries, not a trend towards any single structure. Strongly based artisan systems in the wool and silk industries maintained their structures in face of capitalist competition well into the industrial period. They did, however, eventually succumb to decline, in some cases taking on all the characteristics of a sweated industry as they fought for survival. In the case of silk weaving, the Spitalfields weavers were sweated workers by the end of the Napoleonic Wars, and the repeal of their apprenticeship laws in the Spitalfield Acts. In the case of the Coventry weavers, the compromise of the cottage factory sustained the artisan until the 1860s when new free trade developments sacrificed the industry to the flooding of foreign silk imports. Clothiers in the West Riding woollen industry adopted another compromise in the cooperative or 'company mill' which was viable until the 1850s and 1860s, but they too ultimately succumbed under the pressure of falling rates of profit to mass production in a concentrated factory sector.

Putting-out industries in areas of a strong socioeconomic base and a rising market, or where work processes were more obviously better concentrated, took the road to the factory system. This was the experience of the worsted, linen and cotton industries. Putting-out networks established from early days in areas of population pressure like the East Midlands lent themselves easily, in an industry such as framework knitting, to intensive sweating, as cost-cutting rather than market opportunity determined the constraints on development.

The sweating system and the factory system were the two endpoints which the textile industries reached by the latter half of the nineteenth century. But the forms which industrialization took through the eighteenth and first half of the nineteenth centuries were based on the different routes of the artisan and putting-out systems, as well as, to a relatively limited extent, of the factory system. The forms of work organization which emerged in the nineteenth century cannot, moreover, realistically be termed endpoints, for the structures of work organization continued to shift as industrialization proceeded. The factory system itself was a term which frequently hid more than it revealed. For the size and structures of textile factories in themselves varied enormously, being also subject to constant pressures for change. For an example of this let us now turn to developments in the size of cotton mills in the late eighteenth and early nineteenth centuries.

The size of cotton mills

The size and structure of mills in the cotton industry raise their own problems for the analysis of work organization. We have seen how artisan and putting-out systems took on many different manifestations and remained as viable industrial structures under certain conditions alongside the development of the factory system. But factories too, even within one textile industry such as cotton, developed in a great multiplicity of forms. Extremely small firms fitted in beside the giants. Some were single-process firms; others combined several processes. Some were multistoreyed mills with an assembly-line type of organization. Others were a combination of shacks and work-shops.

We frequently associate the late eighteenth-century cotton industry with new machinery and large factory enterprises. In fact, most of these factories were small, and even the great 'cotton lords' spread their resources over several small, rather than one large factory. Examples of such factories were the well-known, but really rather small establishments of Samuel Old-know and William Ashworth.[70] Oldknow's original mill structure in 1783 for the preparation of yarn cost £90 to expand, con-tained £57 17s 11d worth of machinery, and £261 17s 11d worth of materials. From his mill at Anderton, he employed 59 outweavers. By 1786, he was employing 300 weavers with 500 looms from his Stockport mill and 159 weavers at Anderton; in 1804 he had 550 workers at his Mellor mill. Ashworth, at his New Eagley mill, in 1793 employed 50 operatives spinning and carding.[71] Then there were the giants. M'Connel and Co. in 1795 had an overall capital of £1,769 13s 1½d, and in 1802 employed 312 operatives.[72] In the 1770s and 1780s the Arkwright and Strutt mills were similar to a number of other 'large' establishments in the period. They were multistoreyed buildings with 300–500 workers each and valued at £3,000–£5,000. Each mill had its own water wheel, and the group was enlarged by building similar mills close by.[73] Arkwright's first Nottingham mill in 1772 employed 300 and his Cromford mill employed 200.[74] But by 1783 his second Cromford mill employed 800 and his Manchester mill (in 1780) 600.

There survived, however, a number of little workshops with a carding engine, a few spinning jennies, and hand-, horse-, or rudimentary water-power mechanisms. The room- or floor-letting system used in Manchester and Stockport was common, and one mill in Stockport had twenty-seven masters employing 250 people in total.[75] These were often small businesses which later grew much bigger. Just as often, however, they were second or third mills owned by risk-spreading and diversifying firms. A firm maintaining several mills of varying scale could experiment with new techniques either in its larger factory or in one or two of its smaller mills. Either way, it would avoid the risk of losing everything. Overall averages confirm this picture. As late as 1835, Ure calculated that the average cotton mill employed 175.5 people. Averages in the larger centres in 1816 were 244 for Glasgow, 184.5 for Carlisle, 418 for Stockport, 211.4 for Mansfield, and 115.5 for Preston. In Manchester forty-three mills employed together 12,940. Of these, seven employed less than 100, fourteen employed 100–200, and thirteen 200–400, and only five 400–700.[76]

Concentration of capitals

The increase in the size of cotton mills during the first half of the nineteenth century appeared to be a classic case of the capitalist concentration observed and predicted by Marx. Indeed, Engels in 1844 wrote of this ever increasing concentration of capital in fewer and fewer hands. From the 1830s the small man was being 'squeezed out' as the 'optimum size of firm increased'. But this simplified the picture too much. For in the second quarter of the nineteenth century, capital and credit were available to small as well as to large producers. Even by the 1840s there was still a substantial core of small producers in the industry in Manchester; the average primary process firm in Manchester in 1841 employed 260 hands, and a quarter of all firms employed fewer than 100.[77]

True, large firms could claim economies of scale, but they were just as vulnerable as the small ones to stoppages, insolvency and recession. Before the 1830s new technology was equally accessible to small producers as well as large: small firms took advantage of small steam engines, spinning mules, power looms, water power and traditional building methods.

The self-acting mule in the 1830s, like the water frame in the 1770s and 1780s, was built to the glory of the cotton lords,[78] but in practice most of the self-actors were built on a relatively small scale and they did not come into their own until the 1860s. At the end of the day, 'some giants were as labour intensive as some pygmies, and some pygmies were as power intensive as some giants'.[79]

Our definitions of size are also relative. For the range between small and large was what really mattered for most firms.

Firms employing 150 operatives or less, in terms of shares of the total labour force fell from 28.5 per cent in 1815 to only 12.5 per cent in 1841, while firms employing 500 operatives and above saw their share of the total labour force fall from 44.24 per cent in 1815 to 31.68 in 1841. The minimum efficient size is 151 and the optimum unit was a medium sized firm employing between 151 and 500 operatives.

When the prospects and fate of the moderately sized producers are included, we see that the decline of the small firm was not gradual. It was, perhaps, as rapid as Engels claimed, for if the cotton lords did not take over, moderately sized but definitely larger firms did. There was a rapid switch in the 1820s between small rented units sharing the same factory premises and larger units which occupied a whole factory.[80]

Conclusions

The coexistence of these many different manufacturing structures across the textile industries and even within the cotton industry itself reveals a great deal about the multiple directions of industrialization. There was obviously no through road to the factory system.

The variety in forms of work organization before the main period of industrialization continued afterwards. The forms taken by these structures may have been influenced by the relationships of power and subordination within and between firms. They certainly also owed much to the social context in which work organization developed. Community solidarity established through forms of landholding, long industrial traditions and customs, and the relative supply and militancy of the local labour force certainly did affect in a multitude of ways the nature of original and later manufacturing structures, as well as the scale of production and the local reception of technical change. It is equally true, however, that markets, monopolistic or competitive pressures, profitability and social status all imposed their own discipline on the subsequent directions taken by manufacturing structures. Artisan, putting-out and factory systems were all shaped during the course of industrialization, and all found their own most effective means of responding to the pressure for profits through some combination of increased intensity of work, the division of labour, sweating, mechanization and reorganization. Technologies and the labour forces using them were developed and adapted to these pressures, whether mechanized or not, skilled or unskilled.

CHAPTER 10

The Textile Industries: Technologies

The textile industries form a fascinating terrain for the acting
out of the human drama of technological change. The technologies fundamental to any production process interacted in a
fatal way with the lives of men and women at work, and the
community around them. New technologies meant the pinnacles
of wealth and success to some; destitution to others. The social
divisions they created might appear within one small community, or provoke a great regional split. New technologies were
resisted or welcomed; they were stamped with age and gender as
were the older methods which had gone before. They created
some new jobs, but they also meant unemployment in a whole
new way to many others. New methods, new machines did not
just mean temporary bad times; they could eradicate a trade and
with it the assumption of work for the rest of a person's lifetime.
Along with this, the skills which formed a part of any technology
were developed, cherished, protected and fought for. With this
in view, we turn to the story of technological change in the
eighteenth-century textile industries.

Cloth production from time immemorial relied on the two
basic crafts of spinning and weaving. Both crafts were practised
by enormous numbers of men and women in town and country.
Of the weavers we know something. Their craft was recognized,
organized into guilds or at the least informal associations, and
apprenticeship regulations were enforced. By tradition the
weavers were men, with some capital, for their looms were
pieces of machinery, but women too practised the craft to a
small extent. Of the spinners we know virtually nothing, for
these on the whole were women, working full-time or part-time
for the market and for family consumption. Their labour was
unorganized and unapprenticed; it relied on the dexterity of

their hands and a small tool, the distaff or a simple spinning wheel.

Spinning and weaving both were mechanized over the course of the eighteenth and early nineteenth centuries. We hear much of the ingenious machines contrived to carry out these processes, but little of the implications these machines carried for the skills and livelihoods of their working people. Few historians attempt the difficult task of unravelling the precise social, economic or even political implications of a new technique or change in production method. But a description of production processes and their changes in the eighteenth century is vital to our understanding of most social and artisan political movements then and in the early nineteenth century. One historian has argued this for the case of shipbuilding in the eighteenth century:

> The adze, rasp, clave, auger, chisel, hammer, maul, mallet, mooter, saw . . . will suggest a discussion of the degree of specialization, the dangers, the ownership, the employment of these instruments and hence to the social realities of production . . .[1]

Innovation affected the textile industries long before the eighteenth century. The foot-driven spinning wheel introduced in the sixteenth and seventeenth centuries eventually superseded handspinning and increased productivity by one third. The Dutch loom allowed for four times the previous output, while the knitting frame could produce ten times that of the old hand-knitter. Even the humble flying shuttle doubled productivity. Of course this was small game beside the impressive performance of those spinning frames and mules which by the end of the century had increased productivity by a hundred times.[2]

Spinning

The most striking series of technical changes across the textile industries in the eighteenth century were centred on spinning in the cotton and then the linen industries. Until this time, women spinners were at a premium, something not, however, reflected in their wages. The three machines which revolution-ized the process were Arkwright's water frame, patented in 1769, Hargreaves's jenny, patented in 1770, and subsequently Crompton's mule which came into general use in the 1780s. The invention and early innovation of spinning technology is usually viewed from hindsight as the big break into powered factory production. But if we look historically at these in-ventions, there was no such break. The chronology of their early development shows us techniques developed out of, and adapted to, basic domestic industry. Although, in some cases, like the jenny, the machine was invented in a factory unit, this was fortuitous, for it was first widely used in domestic and workshop settings. The application of power, associating these machines with factory organization, was another step in the process of diffusion. It was definitely a step distinct from the original invention and use of the machines. Let us now turn anew to the chronology of invention in the textile manufacture, with an eye as much to the continuity as to the novelty of this technical change.

The water frame and jenny complemented each other for some time, for the water frame, though it made possible the use of cotton rather than linen for warp as well as weft, still could not produce a uniformly even yarn. It was furthermore, on Arkwright's decision, built on a large scale in water-powered mills. The jenny, however, produced an even yarn, but one too soft to be used for warps, and it was widely worked by hand in the home. The mule, initially also a cottage and hand-powered technology, produced a smooth, strong, fine yarn. All three machines were used until the end of the century, though the jenny was far and away the most important. Colquhoun's estimates of 1789 indicate 310,000 water frame spindles, 700,000 mule spindles, and 1,400,000 jenny spindles. By 1812, however, the mule had taken over with 4,209,570 spindles as against the 310,516 on water frames and only 158,880 on jennies.[3]

The original Hargreaves jenny was too small to use in woollen spinning, but an improved version spread rapidly in the 1770s. The small jenny factories which spread from cotton into wool had a substantial impact on output, but an ambiguous reception. Where in 1715 seven combers and twenty-five weavers kept two hundred and fifty worsted spinners employed, hand jennies reduced the weaving–spinning ratio to one weaver to four spinners. In Yorkshire both the jenny and the carding engine were introduced by domestic spinners. The jenny which reached the Holmfirth district of Yorkshire in 1776 'contained 18 spindles and was hailed as a prodigy'.

> Every weaver learned to spin on the jenny, every clothier (or manufacturer) had one or more in his house, and also kept a number of women spinning yarn for him in their cottages.

The jenny came into general use in Leeds in the 1780s, and by 1793 Benjamin Gott had three or four dozen of these in his large mill at Bean Ing. By this time, however, objections were being raised about the jenny's impact on employment, and in 1806 a group of merchants and master clothiers in Saddleworth met to pass a resolution opposing the factory system in woollen spinning and weaving.[4] In the West Country, until the 1790s, it was only on the fringes of the industry that such machinery was accepted. This was an area which, unlike Yorkshire, was not expanding, and spinning occupied women over a wide agricultural area which offered no alternative employment. The jenny was, on the whole, accepted more easily in the North, and did not pose such a big threat to employment as in the South, for many in the North spun worsted, and the machine was not used for this.[5]

The new spinning technology centred on the jenny was developed within the context of domestic, small workshop and protofactory production. There was no striking and revolutionary break between mechanized factory technology and cottage manufacture. The jenny could, of course, be worked in the home and was still run by women.[6] It was, however, usually jennies of less than twenty spindles which were used at home, and the technique was widely used on this scale. And this was so, in spite of the fact that Hargreaves had actually developed his improved jenny within a factory in Hockley near Nottingham. This factory was owned by Hargreaves's partner Thomas James, and though the jenny was cheaper to build and simpler in design than the cottage stocking frame, it was initially kept within a mill which was operated as a 'safe box'. Only trusted workmen were admitted to it, and there were no windows on the bottom floor.

During the 1770s the James-Hargreaves mill was rival to Arkwright's mill. The jennies were also quickly absorbed into the Nottingham hosiers' workshops and warehouses. These hosiers' business premises

> consisted of a few rooms, or perhaps the whole of a substantial house near the centre of Nottingham. The first distinct warehouses were being built by the 'sixties. Apart from accommodation for stock and keeping accounts, the hosiers' premises contained some machinery: several stocking frames to supply urgent or special demands, twisting mills, warping mills and sometimes silk-throwing machines, no doubt based on Lombe's principle. In the country districts, a handful of hosiers also operated fulling mills. It was not difficult to add another room to accommodate a few jennies.[7]

But the jenny which spread around Lancashire at the time and later in the woollen districts was a cottage and a small factory technology. Larger jennies were used in the so-called jenny factories. These jennies became linked to machine carding, and it was the small factories containing these machines which spread so rapidly at the time.

Just why the jenny was taken up in the cotton manufacture raises a series of questions about the connections between the organization of the industry and technical change. Stephen Marglin argued that there were no special technical reasons for the factory system; the factory system simply guaranteed to the manufacturer a higher proportion of the surplus and any gains to be had through increased labour productivity. We can take the opposite line, and ask if there was anything inherent in the pre-factory organization of the cotton industry to encourage the introduction of new techniques. The answer to that is yes, but the reasons are not, at first sight, obvious. First, the putting-out system itself did not create any incentives for the introduction of the spinning jenny. For merchants and factors did not control the intensity or efficiency of their labour unless they controlled markets and prices. In the Lancashire cotton industry there was no all-pervasive industrial concentration. Competitive prices were imposed on merchants as well as upon workers; the worker had outlets for his yarn other than any individual factor or merchant. In this setting the merchants had no special price incentive to introduce the spinning jenny, though they did stand to gain from higher and readier supplies of yarn. The ones who stood to gain most were those who owned and ran the spinning jennies themselves. It was the cottage producers and those who ran small centralized workshops who reaped the first gains in efficiency from the jenny, and they did so until the merchants and factors saw the gains to be had through setting up their own larger jenny factories. It was not the putting-out system itself which brought the jenny to the cotton manufacture, but a dispersed industrial structure which created opportunities for the direct producers to reap some of the gains from increasing their efficiency.

The mule was only a development on the jenny, and it too was initially used at home and in small factories. The improvement made was simply the addition of roller drafting to the spinning jenny. This could have been applied earlier and equally successfully to the jenny, but the jenny was not developed in its own right.[8] The mule was also used initially on a putting-out basis. It was, however, more expensive than the jenny, and though it remained an integral part of the domestic system until the 1790s, it was also quickly moved into mule shed factories. Still, these were small enterprises when compared with Arkwright's warp-spinning mills.[9] Until the late 1780s mules were only built on a scale up to 144 spindles, and each mule was manually operated one at a time. It was the application of water power in 1790 which substantially increased productivity, for this allowed the installation of mules in pairs, one in front of the spinner and one behind. Steam power was applied in the same way later in the 1790s.[10]

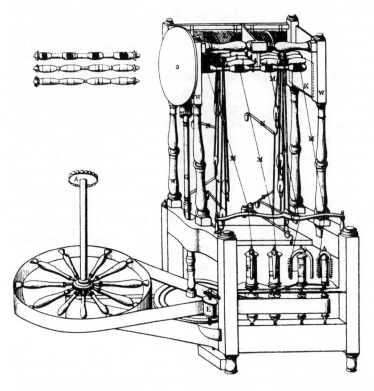

The Arkwright water frame, patented 1769. An early design with four spindles (Andrew Ure, *The Cotton Manufacture of Great Britain*, 1836)

Machine spinning came much more slowly to the woollen industry than it did to either cotton or worsted, largely because of important basic differences in materials and necessary technical processes. Woollen goods were made from short-fibred wool, worsted from long; and wool was carded while worsted and cotton were combed. The object of carding was to combine all the fibres so they would interlock or 'felt', while combing was meant to lay all the fibres in the same direction. The upshot of these different materials and preparatory processes was to affect spinning technology. The object in spinning both worsted and cotton was to elongate and to stretch fibres straight, thereby continuing the combing and roving processes; in the woollen manufacture, the object was not to straighten the fibres, but to preserve their natural curl, while also laying these fibres in the direction of the length of the thread. Spinning was not to impair the felting process started in the carding of the wool. The mule was eventually adapted to this task, but even in the early decades of the nineteenth century

> the mule has not, till lately, been in much repute for spinning woollen yarn, and the jenny is still thought to spin better yarn; but we have no doubt that when certain modifications are made, it will become a much more perfect method than the jenny, being much less dependent on the discretion and dexterity of the spinner.[11]

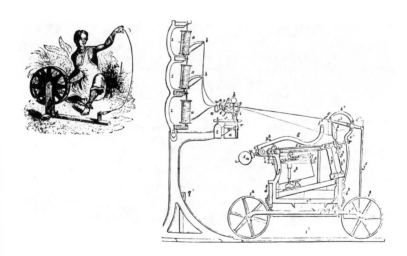

'Primitive' Indian spinner and 'modern' spinning mule
(Andrew Ure, *Dictionary of Arts, Manufactures and Mines*, 1846)

We thus see that there were good technical reasons for the different receptions offered to new spinning techniques. The jenny and, later, the mule were much greater threats to woollen spinners than they were to worsted spinners. But the worsted spinners faced the much greater hiatus of Arkwright's water frame, for after the combing process, the processes of the worsted and cotton manufacture were very similar. Arkwright's water frame, developed for the cotton manufacture, was quickly adapted to worsteds.

Roller or water frame spinning, invented by Wyatt in 1738 and brought into use by Arkwright, had a revolutionary impact on all three processes of carding, roving and spinning. The water frame could carry out all these processes. It was, furthermore, driven by power long before the mule was used with anything but handpower. The frame could also produce the necessary twist for warps of superior strength and wiry smoothness; it was used to make water twist of low counts cheaper than mule twist. This was a big advantage for power-loom weaving,

Spinning mule
(Edward Baines, *History of the Cotton Manufacture of Great Britain,* 1835)

and so the technique, in spite of other defects, was further developed to become the efficient and widely practised process of throstle spinning.

Arkwright's water frame has been virtually identified with the origins of the factory system. It has been assumed that this was a machine which required a large capital outlay, a source of power and centralized production. But in fact it was originally designed as a small machine turned by hand and capable of being used in the home. It was Arkwright's patent which enclosed the machine within a factory, had it built only to large-scale specifications, and henceforth refused the use of it to anyone without a thousand-spindle mill. A model of the original water frame is in the Science Museum of London. It

> spins beautifully and shows that the water frame could have been built in small units, placed in cottages and turned by hand. In other words, it could have been used like the jenny as a domestic spinning machine. One member of the Arkwright's partnership, I suspect it was Arkwright himself, for it seems in character, must have realized that if this had happened they would have lost control of the patent, for everyone would have copied it and built their own machine in the privacy of their own homes. By restricting the licences to units of a thousand spindles, it became economic only when they were erected in a water-powered mill. This was a vital decision in the development of the textile industry and of the Industrial Revolution which never seems to have been recognized before.[12]

But Arkwright's 100 per cent rate of profit led all to believe that the machinery and the factory system were as one. The jenny and the mule were still popular, however, and continued to be developed in their own right, for the water frame could not spin high-quality yarn or the fine weft; and machinery was soon developed which could take over the hand preparatory processes for jenny and mule spinning. Combing machinery for the worsted industry and the billy which carded cotton into rovings gave renewed advantage to the mule and the jenny.[13] And the mule, easily adapted to power, soon posed a threat to large country warp-spinning mills.

In spite of the association generally assumed between the development of motive power and the primary innovations of the textile industry, it was actually the case that steam power had a very belated effect. As von Tunzelman has argued, even in the precocious cotton industry all the technological breakthroughs in cotton spinning were originally developed for other forms of power – hand, horse and water. The application of steam power on any sizeable scale can be charted only from the last years of the eighteenth century. By 1800 at the most 1,500–2,000 horse power in the form of steam engines was used to turn textile machinery and accounted for up to one quarter of the cotton processed in that year.[14] Hand-operated machinery was still common in carding and spinning until well after 1815. The mule was hand-worked until the 1790s, and the handloom dominated weaving until after 1815. No power was applied in bleaching and printing until the later 1790s.[15]

The spinning of silk, unlike that of other textiles, was mechanized and centralized from the beginning of the eighteenth century. Some of the stages for making the organzine used for the silk warp were amenable to mechanization. The various stages involved winding the silk from the skeins upon bobbins in winding machines; sorting it into different qualities; spinning and twisting each individual thread in a mill; bringing together on fresh bobbins two or more threads already spun or twisted; twisting two or more threads together by means of the mill; and sorting the skein of twist or organzine according to fineness. The winding and twisting stages were both mechanized. The first was done on a bevelled wheel and attended to by children.

> The constant attendance of children upon this winding machine is requisite, in order to join the ends of any threads which may be broken in winding, and when the skeins are exhausted, to place new ones upon the swifts.[16]

The twisting was done on the throwing mill. Large twisting mills on Thomas Lombe's design contained machines in a circular frame about 15 feet in diameter. But the machine was basically a large version of a hand technique.

> The great mills for twisting silk, originally introduced by Messrs. Lombe, though very complicated, are simple in their operations, because the complexity arises from the great number of spindles which are actuated by the same movement, every one of which produces its effect independent of the other . . .

A similar small version of the machine was used with only thirteen spindles as opposed to the multiples of eighty-four in the large silk mills. The small-scale machine was

> intended to be turned by hand, a method which is too expensive for this country, but is common in the south of France, where many artisans purchase their silk in the raw state, and employ their wives or children to prepare it by these machines . . .[17]

This chronology of the early development and diffusion of the new spinning technology shows us a much closer integration with rural manufacture and artisan or domestic industry than we generally perceive. In the face of resistance to new technology and the pull of the older methods, the obvious question is why the new machines spread in the first place. When we turn to the standard economic cost-benefit analysis, it seems unlikely there was much push to the new technology from wage rates, for spinners were notoriously ill-paid. Fixed capital costs were also unlikely to provide any incentive. But labour costs, rather than wage rates, and capital costs, including circulating capital, probably did provide a factor, for labour costs were increased by embezzlement and poor quality; circulating capital costs by delays, costs of intermediaries and the uncertainties of supply. But the much more positive inducement was increases in labour productivity. The gains to be had by simply incorporating a relatively inexpensive machine with x number of spindles into a household which formerly had only one or two one-spindle wheels were obvious.

But it is also clear that the opportunities for such gains in productivity within the old framework of domestic industry were not really exploited beyond a very short period. Their potential for large-scale production was obvious, and factors, merchants and substantial manufacturers soon adapted their warehouses and workshops to take the new machines. Productivity gains in cotton became increasingly associated with scale and factory organization, and the pull of such productivity gains provided the prime mover of innovation; the prospect of windfall gains, not the spectre of falling profit margins, lured cotton manufacturers into changing their methods.

The effect of the new spinning techniques on employment and on the division of labour varied enormously among regions and branches of the textile industry. We will look at these effects in comparison with those of other processes later in the chapter. Now we turn to innovation in the weaving and finishing processes.

Weaving

Few improvements were made in weaving, apart from the flying shuttle, until the introduction of the power loom. The power loom was used to a certain extent in cotton at the end of the eighteenth century, but did not spread to wool and worsteds until the early to mid-nineteenth. The flying shuttle, however, had an important impact on the productivity of existing looms. By the 1770s there existed a broad and a narrow loom. The broad loom made cloth up to a hundred inches wide and needed two operators, while the narrow loom made up to half that width. The flying shuttle both increased productivity and in broad cloth manufacture did away with one of the operators. Its other major contribution was to make it common to weave woollen cloth with a design.[18] Though introduced in 1733, the improvement was not widely used until the 1760s and 1770s, and in fact met resistance in East Anglia and Lancashire. It was, however, well received in Yorkshire where it eased the weaving of broad cloth.[19]

The really important weaving innovations of the eighteenth and early nineteenth centuries affected the silk manufacture first. Even here, weaving was not mechanized until well into the nineteenth century for largely technical reasons, for applying power to silk weaving would not save a great deal of labour. The weaver could not attend several looms at once and under any conditions would have to spend a great deal of time on 'picking the porry', that is, removing the roughness and inequalities in the warp threads. Handweaving, however, involved varying degrees of skill depending on the fineness and design of the product.

> Plain weaving is thus seen to be a very simple operation. A certain degree of proficiency in the art may doubtless be quickly and easily attained, but much practice and attention are nevertheless required, in order to form a dexterous weaver, so as to enable him to produce well woven fabrics, and to accomplish within a given time such a portion of work as will earn for him a competent subsistence.[20]

Improvements were made in figured weaving in the early nineteenth century with the introduction of the 'drawboy' in 1807 to replace one of the weavers on the complicated drawloom. And by the 1820s the Jacquard loom had been introduced from France. The Jacquard, however, was initially popular only in the Spitalfields silk industry, and the woollen industry was slow to recognize the attributes of the new machine. The Jacquard spread first into the Scottish carpet manufacture and thence by 1822 into the Coventry ribbon-weaving industry. It became part of the West Riding woollen industry only in 1827.[21]

Ribbon weaving was first done on a single handloom making one ribbon at a time; the Dutch loom, which wove several ribbons at once, was introduced in 1770, but was widely used only from the early nineteenth century. The Dutch engine loom, in spite of its name, was worked by the hands with treadles for the feet like a common loom, but each warp occupied a separate shuttle, and the shuttles were impelled by an instrument called a ladder. Male and female journey hands who had completed a five- to seven-year apprenticeship were traditionally employed on the single handlooms, but the new Dutch looms were generally worked by skilled men.[22] The only strike heard of before the end of the Napoleonic Wars in the Coventry trade was occasioned by a journey hand who wished his wife to be employed on a Dutch loom. The strike was successful, and over this period no woman was allowed to use the Dutch loom. In the period after 1815, when the hold of the larger manufacturers was challenged by small capitalists employing cheap labour, women were put on to the machines. The larger manufacturers only survived by increasing mechanization, introducing silk throwing on their premises and putting up handloom factories containing improved engine looms.[23] But in the Spitalfields silk industry and the West Country woollen industry, women did use the Dutch or engine loom. Attempts were made to exclude them from the use of this more advanced technology, but always at times of high unemployment and crisis in the industries.

Powered weaving was slow to come, not only to silk and wool, but even to cotton and linen. For the power loom took a long time to perfect. Patented in 1789, it was not until 1813 that it assumed its traditional form, and it was still defective by 1833.[24] Its development depended upon some general considerations such as the nature and property of the raw materials from which the yarns were manufactured, the qualities of yarn to be woven, the system of preparation of warp and weft for the yarn, and the organization of industry and attitudes of operatives to the consequences of the new techniques.[25] But basic mechanical imperfections still plagued the machine. The necessity of dressing the webs from time to time after they were put into the looms made it impossible for one person to do more than attend to one loom. By 1813, however, Horrocks's improved power loom was combined with Radcliffe's dressing frame. Subsequently, some enthusiastic contemporaries estimated that a boy or girl could attend to two looms and produce three times as much as the best handweaver.

The slow introduction of the power loom was to a large extent a problem of fitting this technology to other processes mechanized earlier. The power loom produced for only low-quality markets. The handloom continued to dominate the fine muslin trade, but was also a formidable rival for plain and coarse goods because of the early technical difficulties of the power loom.

In 1813 there were not more than 100 dressing machines and 2,400 power looms in use, and there were approximately 240,000 handlooms.[26] Productivity differentials were difficult to assess. In 1819 there was one weaver to one or two looms, but the average ratio of power to hand output ranged between two to one and three to one. By 1829 the number of handlooms had fallen to 225,000,[27] and power looms had increased to 55,000. But output per handloom had increased by 25–30 per cent. This was probably the result of a combination of two factors, the application of more labour to each handloom and technical improvements. The dandy loom, one such technical improvement, was a species of handloom operated by a combination of mechanical and hand movements; it was alleged to keep pace with the power loom.[28] Between 1819 and 1842, however, the average speed of the power loom had increased from 60 to 140 picks per minute. In 1820 a good handweaver could weave 172,000 picks per week, and a power-loom weaver with two looms could weave as much as 604,800 picks per week. In 1825 a power weaver with two looms and no assistance could weave 1,000,000 picks per week. By 1835 a weaver could attend four looms with the assistance of a tenter and weave 1,759,000 picks per week.[29]

Finishing techniques

Perhaps the most celebrated improvements in the eighteenth-century textile industries were in the finishing processes – cropping and shearing in the woollen manufacture. For it was the artisans from these trades who, along with some handloom weavers, framework knitters and worsted combers, led the celebrated Luddite machine-breaking episodes of the early years of the nineteenth century.

The mechanization of finishing processes was conducted

Gig mill
(*The Useful Arts and Manufactures of Great Britain,* n.d.)

against a background of urban skilled workers with similar characteristics. The shearman worked not at home, but in finishing shops, generally owned by independent master dressers. The first innovation, the gig mill, mechanized the process of raising the nap. Where it formerly took one man 88–100 hours to raise one piece by hand, with a machine one man and two boys could do the same job in 12 hours. Once the nap on the cloth was raised, all the surface wool had then to be removed by a process called shearing or cropping. The original shears were two large, flat, steel blades bent into a circular bow. These were worked by two shearmen whose skill consisted in operating them in a regular and parallel motion so that every part of the surface was equally cropped. Towards the end of the century a shearing frame was introduced, which simply fitted the shears into a frame moved by means of a carriage over the cloth. The machine was not very effective and in fact still required 'great care and attention to make the different cuttings

Broad perpetual shearing frame
(*The Useful Arts and Manufactures of Great Britain,* n.d.)

join, in order to cut equally over the whole surface'.[30] The frame was used, however, and was improved by the early nineteenth century into a perpetual shearing machine, and then into a rotary machine.

A similar spirit of artisan skill and militancy existed among the cotton calico printers. But their struggle with mechanization was deflected by geographical division. The skilled calico printers centred in London, who refused to use the new cylinder printing processes, were by-passed by manufacturers such as Peel who established new centres of calico printing in Lancashire. But skilled calico printers there too soon grew proud and militant, and calico printing was moved to the countryside where new techniques were invented to tap a largely female labour force.

The impact of textile machinery

This exposition of the course of technical change in textiles opens up a series of issues only infrequently raised by economic historians. The main big issue concerns the effect which the new technologies had on employment, skills and the division of labour. Secondly one needs to ask how closely linked changes in work organization were with these new technologies.

We have no means of quantifying the effects of spinning innovations on employment. To some extent we can gauge the effects through the regions and industries which experienced high levels of resistance to the new machinery. But this is not enough by itself, for resistance was a political act requiring consciousness and hope. Technological unemployment could also entail a sense of hopelessness and passive acceptance. There was, in the early days of the jenny at least, less distress among the cotton spinners than the woollen spinners. Cotton was more localized, and the spinners had other branches of the manufacture to turn to in a region of rapidly growing trade. Wool, on the other hand, was diffused throughout the country, and the lot of spinners in declining agricultural districts was much more difficult.[31]

The economic theory of technological change and the economic historians who write within its framework rarely have any explanation to offer for the waves of resistance to machinery which swept through the textile industries in the eighteenth century and in the early nineteenth. The orthodox economist usually assumes, if he cannot prove, that technological change creates more, not less, employment; hence the actions of Luddites and antimachinery protests of workers are seen to be irrational or at least mistaken, or are taken to be a means of expressing other demands.

Economists certainly recognize 'factor bias' in innovation, that is, that innovations could be more or less labour-intensive; but most agree to a favourable overall economic impact. Few, apart from J. R. Hicks, accede to the possibility and even the likelihood first discussed by Ricardo, that rapid technical change in the first part of the nineteenth century may well have called for the transfer of resources from circulating capital (or wage goods) to fixed capital, and so simultaneously reduced overall employment while raising profits.[32] Did this happen in the Industrial Revolution? Economic historians only reply with very tentative empirical estimates of aggregative economic categories – contributions of capital, labour and the 'residual' (the catchall that includes technological change). The more careful among them at least concede to the dubious conclusions drawn from such aggregate data, and counsel the study of individual innovations.

But on individual innovations they also prefer economic orthodoxy. Von Tunzelman, summing up his survey of a number of innovations, plumps for long-term considerations of profitability rather than short-term cost-cutting as the inducement to most eighteenth-century innovation. Where labour appeared to be adversely affected by technical change, he argues, it was largely in those industries which had been bypassed by new technology. The Yorkshire cloth dressers, he concedes, did have a quarrel with machinery, but they were a small minority.[33]

Conclusions like this are all, however, based on the cotton industry, not the woollen, on the experience of the North, and not of the South. They give a misleading picture of the impact of technological change, one which consigns to a netherworld all the spinners, weavers and cloth finishers of the much larger eighteenth-century woollen industry. And large declining industrial regions of the South and the Midlands, and even agricultural regions where women's spinning, knitting and lacemaking formed a substantial part of the local economy, have no part in our current optimistic histories of technological change. In the textile industry alone, the spinners in the eighteenth century and the handweavers in the nineteenth were the majority. Spinning and weaving machinery did substitute directly for their labour, and though we may well argue that hand processes continued, they did so in competition with machinery, and so at lower wages and more intensive work.

It is fashionable now for economic historians and even social historians to argue that most technology was really labour-using, not labour-saving. Indeed, this book too stresses the labour intensity of many processes. But the upshot of the argument should *not* lie with dismissing the existence of a machinery question either in the eighteenth century or in the nineteenth. For the high levels of resistance to new technology in older or declining industries, and even in the new cotton industry with its meteoric growth path, indicate that many were losing jobs or suffering wage reductions in the major manufacturing industry in Britain at the time.

Some might point to inconsistency in such diverse claims. But they are those who prefer aggregative unilinear analysis. The historian with some sensitivity to the differences over industries, regions, the economic cycle and the labour force finds no such inconsistency. Many did lose their jobs with the coming of technological innovation, but the labour of many others also became more intensive. The fixed capital of textile entrepreneurs may have increased only slowly, but the 'capital' of skilled workers – their traditional crafts and skills – were being undermined by dilution.

It is not enough, as Habakkuk himself argued, to point to the labour intensity of British industry. What is equally at issue is the skill intensity of this industry. And any effect of new spinning or weaving technology must depend on *whose* employment was reduced and whose increased; it must depend on the impact on the division of labour.

New techniques, skills, and the division of labour

Though much of the new textile technology was established in continuity with handworking and domestic production, it did change the sexual division of labour. The original jenny was best suited to being worked by children. The small country jennies of about twelve spindles had a horizontal wheel and a treadle 'which required an awkward and constrained posture'.

> The awkward posture required to spin on them was discouraging to grown up people, while they saw with a degree

of surprise, children from nine to twelve years of age, manage them with dexterity, which brought plenty into families that were before overburthened with children.[34]

But a vertical wheel was substituted for the horizontal, and the treadle was replaced by a simple contrivance managed by the hand. This improved jenny, generally equipped with sixty to eighty spindles, spread widely in the eighteenth century, and was mainly operated by women. The water frame, similarly, was run by children and girls. Larger jennies of up to 120 spindles and large handmules were built in the 1790s, but these needed male operatives, and, as men cost more than half as much again as women to employ, this was not a popular alternative. 'Thus so long as low wage female labour could easily be found there was not much temptation to search for other sources of power.'[35]

Many spinners took readily to the new jennies, for these were small enough to use at home and helped enormously to increase the output of yarn. For others, who could not acquire the new jennies, there were lower wages, but also the alternative of more employment in carding and preparing wool. The high or at least decent wages of some early jenny spinners soon attracted more workers, and wages fell. By 1780 there were complaints that 'women who had been earning 8s to 9s a week on jennies of 24 spindles' could now earn only from 4s to 6s a week. And bigger jennies with more than eighty spindles proved a much more substantial threat. Larger jennies took away women's jobs and were furthermore located outside the home in the new jenny factories. The early jennies were a part of the domestic system, the machines of the poor; the large ones were monopolized by capitalists. As a contemporary petition put it,

> that the Jenneys are in the Hands of the Poor, and the Patent Machines are generally in the Hands of the Rich; and that the work is better manufactured by small Jenneys than by large ones.[36]

It was during its domestic industry phase that mulespinning too became a male occupation and acquired craft status. When the mule was still manually operated considerable strength was needed for pushing the carriage back and forth. After power was applied women could be used, but substantial physical stamina was still required in order to maintain a given pace of work with coordination and attentiveness for twelve to thirteen hours a day. The mule also required considerably more skill than the jenny.

> To one accustomed to operating the jenny as Crompton was, the spinning on his new machine would be a work of art, a display of skill and judgement.[37]

The application of power to the mule did not mean automatic working: it only turned the rollers and drove the carriage and spindles during the spinning action. The operative still had to revolve the spindles and guide the spun yarn on the spindles in the form of a cop by the use of his hands as the carriage made its inward run. The spinner had to coordinate three operations simultaneously. First he had to push the carriage towards the roller beam. Secondly, by means of a faller shaft and wire he had to control the winding of yarn on to a cop chase so as to form a correctly shaped cop. Thirdly he had to turn the spindles by means of the fly so as to wind up the yarn being released by the inward motion of the spindles. He had to avoid breaking the yarn, yet never permit it to go slack. After forty years of rigorous development along a number of different avenues, the spinning mule still required the continued attendance of a skilled operative.[38] Other factors which, from its early domestic phase, helped to restrict the machine to a predominantly male workforce were the substantial amounts of capital required to buy or build it, and the skill to maintain it. Still, women could and did learn the skills and were quite widely used even into the 1830s on the smaller mules.[39]

The possible threat posed by the more widespread introduction of women in mulespinning was reinforced by manufacturers' search for a self-acting mule. The spinners after 1824 formed a close-knit union. But an earlier existence of the union manifested itself in 1810 and 1818 in strikes over piece rates and women workers. The membership of the union in Glasgow alone was estimated at 800, 'enough to exercise direct control over several thousand dependent kin in factories'.[40] The control exercised by these spinners was what prompted a consortium of mill owners to seek the invention of a self-acting mule finally achieved in 1830 by Richard Roberts. Roberts's improvement was a new design of headstock which incorporated a system of closed loop feedback control. But in spite of manufacturers' hopes of using cheaper labour, the spinner still retained his tradesman status. His workload was lighter, but it was usual for him to continue to make manual adjustments throughout the course of each set of cops spun. He also had important social attributes (to be explored below) which still made it advantageous for employers to continue with their traditional staffing arrangements. The self-actors could, however, be made with more spindles, and unions were prepared to accept these along with speed-up as the price of retaining the spinners' tradesman status.[41]

At the end of the eighteenth century spinning and weaving employed both skilled male tradesmen and women. As spinning passed out of the hands of female domestic workers, those in the linen trade turned increasingly to handweaving.[42] In the cotton industry, the warp-spinning mills also generated their own demand for less skilled weavers – the aged, women and children, and Irish migrants who wove the coarser grades of cloth. It was their employment which was threatened initially by the introduction of the power loom. The weaving of fine goods still primarily remained the preserve of the male craftworker. The skill of the Scottish linen weavers was turned to the production of fine muslins – several thousand looms in the Glasgow area which had formerly produced linen, silks, cambrics and lawns were changed over in the last decades of the eighteenth century to fine cottons, and the power loom was slow to spread in the area thanks to its unsuitability to the finer counts of yarn.[43] Such fine weaving was in some cases incorporated into the factory, but as skilled work on specially developed hand machines. Oldknow, for example, introduced the manufacture of figured muslins, but these were made on newly invented costly looms called drawing engines, worked by men assisted by boys.[44]

Specialization among products in weaving implied in effect a choice between factory production and domestic production. Until the mid-nineteenth century, production of fine muslins meant handlooms. This is borne out in a statement made by Kirkman Finlay before the Select Committee on Manufactures, Commerce and Shipping in 1833.

> A grand mistake exists in supposing that the power loom supplants the hand loom universally: the power loom used in Scotland manufactures a kind of goods in general, which the hand loom weaver of Scotland was not in the practice of working at all. . . . Before power loom weaving was introduced at all in Scotland about the year 1814–15, the kind of goods generally manufactured by them were not manufactured at all in Scotland . . . I would also say that the hand loom weaver can work a great many things which it would not be the interest of any power loom manufacturer to make, especially all the finest goods, fancy goods of all kinds . . . it can never be the interest of any power loom manufacturer to make a kind of goods of which he cannot regularly and constantly dispose of a large quantity of the same kind.[45]

Baines, too, in the context of explaining wage differences, emphasized:

> There is . . . a distinction to be made among hand loom weavers according to the kind of goods on which they are employed. Those employed in weaving fancy articles, which require skill and care . . . obtain much better wages than the weaver of plain goods . . . which require very little strength or care.[46]

Defects in the power loom as late as 1828 affecting the quality of the cloth mainly arose from a need for regular motion to draw the cloth forward as it was woven and for a simple means of varying the rate of take-up for the different qualities of fabric.[47]

The diffusion of the power loom was effectively a phenomenon of the 1820s and 1830s, as is indicated in these levels of employment. By 1788, 60,000 were employed in factory spinning, practically none in power weaving, and 108,000 in handloom weaving. By 1806 the number of factory spinners had increased to 90,000 and the number of handloom weavers to 184,000. By 1824 there were 122,000 factory spinners, 45,000 power weavers, and 210,000 hand weavers. Those employed in the factory multiplied again by 1833 to 133,000 in spinning, 73,000 in power weaving, and 213,000 in hand weaving.[48]

The general characteristics of labour employed in each process and the diversity of wage levels are indicated in the following table:

Table 14. *Structure of employment and wages by process, 1833*[49]

TECHNIQUE AND EMPLOYEE	WAGES	
Male and female adults involved in cleaning and spreading	8s	3d
Adult male carders and overlookers	23s	6d
Female tenters	8s	
Overlooking spinners (male adults)	29s	3d
Spinners (male and female adults)	25s	8d
Piecers (children)	5s	4½d
Overlooking throstle spinners (male adults)	22s	4½d
Warpers (male and female adults)	12s	3d
Weavers (females)	10s	10d
Dressers (male adults)	27s	9½d
Reelers (female adults and children)	7s	11½d
Engineers, firemen, mechanics	20s	6d

There was, of course, widespread child labour in the early factories. Children entered the factories in a technical ancillary capacity. James M'Nish testified before the Parliamentary Committee of 1831:

They can only be of a certain size and under a certain age; they have to go under the threads to wipe down the machinery; if they are too large they break the threads and destroy the work, and were they large they could not roll below the carriages; as the machinery is at present constructed, only children 9–11 can do the work as it ought to be done.[50]

Most of these children came as assistants to adult spinners on a subcontract basis, that is, spinners were paid by the piece and hired their own assistants. In power-loom weaving, too, the tenters were often paid by the weavers. In 1816, 54 per cent of employees under eighteen in thirteen mills around Preston and 59 per cent of those in eleven mills in the country around the town were paid by spinners and rovers. 'Thirty years ago they were almost all children and at least five times as many under 10 years of age as at present.'[51] It was also noticed that persons employed in water spinning were older than previous employees had been. 'Such young children cannot so well be employed in water spinning as they could formerly and as they can I understand be employed in mule spinning.'[52]

Child labour was also widespread in the skilled crafts. A case was calico printing. Here the use of children and young girls was so lucrative that it provided the basis for an alternative path of technological development. The traditional technique of block printing survived into the mid-nineteenth century in spite of the development of printing machinery. Children were widely employed in this process as 'teerers', spreading the liquid colour evenly on a floating sieve with a small handbrush. Children from six years of age were employed in this and other repetitive work in a highly seasonable trade which frequently required fourteen- to sixteen-hour days and nightwork.[53] The calico printers also employed a large number of girls. And indeed the labour-intensive 'block pinning' and 'pencilling' processes introduced in the Lancashire industry tapped a female labour force.[54]

Resistance to new technology

Not only were technologies associated with specific social and gender qualities, but there were widespread differences in their reception in workplace and community. Resistance to the introduction of new technologies was not confined to the factory age; it was also endemic to the early stages of industrialization. We have seen in Chapter 5 that such resistance was part of the story of industrial decline. But the textile industries experienced widespread differences in their patterns of technological diffusion and labour resistance. Power sources, product choice, employment structure and community relations all affected the extent to which a region took up, resisted or ignored any particular technical change. In some areas such as Essex the reaction to technical change was variable. The weavers' there revolted against the factory system, but accepted the flying shuttle.[55] In Yorkshire the jenny and carding engine were introduced by domestic spinners in times of expanding employment, but combing machinery was resisted in the early nineteenth century. Spinning machinery was widely resisted in the West Country, and although in spite of this jennies were introduced over a wide area by the 1790s. And as noted before, Wiltshire and Somerset mounted fiercer resistance to finishing machinery than did Gloucestershire.[56] The prospect of such machinery arising in Wiltshire in fact lay behind the Wiltshire Outrages of 1802, and the shearmen received widespread public support from weavers who feared that they themselves would soon be resisting factory weaving, from small master dressers who feared the competition of large machine owners, and from some large clothiers who had other reasons for disliking the innovators.[57] The flying shuttle met a good reception in Yorkshire, but was widely resisted in East Anglia, Lancashire and the West Country. Even as late as 1822, an attempt to enforce the use of the flying shuttle along with a reduction in weaving wage rates led to widespread riots.[58]

The regional differences in the reception of new technology at least in the woollen and worsted industries have been attributed to the extent of social polarization. It is argued that the sharp polarity of master and man in the South contrasted with the socially more uniform small weaver communities of the North. This seems to have manifested itself in more forceful workers' resistance to machinery, instances of which we have noted already. In Bradford-on-Avon in 1791 the mob destroyed an advanced scribbling machine. In Trowbridge in 1785–and between 1810 and 1813 workpeople rioted against the flying shuttle, as they did in west Wiltshire, postponing its introduction until 1816. Somerset clothiers who wanted their cloth dressed by the gig mill had to send it as far away as ninety miles.[59] Yet it is also true that protest against machinery was particularly strong in Lancashire, Yorkshire and the Midlands. Hargreaves's first spinning jenny was destroyed by a mob in 1767, and in 1769 more machines were destroyed at Turton, Bolton and Bury. A number of weavers were involved in these and later disturbances, and a contemporary wrote that the output of weft from jennies

> gave uneasiness to the country people, and the weavers were afraid lest the manufacturers should demand finer weft woven at the former prices, which occasioned some risings, and the jennies were opposed . . .

And in 1779 there the celebrated jenny riots around Blackburn.[60]

Hargreaves's move to the Midlands was not inspired by the prospect of a more docile labour force, for Nottingham in the second half of the eighteenth century had a well-known reputation for popular protest, riot and attacks on machinery,[61] all of which probably stemmed from the marked social divisions within the hosiery industry. Cloth-finishing machinery, stocking frames and power looms became the targets of disciplined Luddite attacks in 1811–12 across the West Riding, Nottinghamshire and Lancashire.

The Luddite movement was by and large based on the small local community and not on urban social structures. The personal, kinship and other social connections within a workshop culture or small quasi-peasant community created close bonds around the saboteurs. They were the necessary cover of secrecy for the highly disciplined 'guerrilla bands' moving from village to village at night that characterized Luddism in Nottinghamshire, Derbyshire and Leicestershire. Thompson in fact describes this Luddism as 'the nearest thing to a peasants' revolt of industrial workers'. The Luddites 'knew no national leadership or policy they could trust or identify with', and Luddism was always strongest in 'the local community and most coherent when engaged in limited industrial actions'.[62]

The act of machine-breaking, as we have seen, had a long history and was bound up in various degrees with the preservation of domestic industry, employment, and, especially in the case of the framework knitters, with the preservation of the customs, standards of quality and skill levels of the trade. In the case of framework knitting it was not the machine as such which provoked hostility, but the new malpractices of employers over apprenticeship and quality. This resistance to machinery, manifested in the form of machine-breaking and violent disputes, continued in the textile industries right through the 1830s, with particularly explosive episodes in Lancashire and Yorkshire in 1826 and 1829, and further bitter disputes in Scotland over self-acting mules throughout the following decade.

CHAPTER 11

The Metal and Hardware Trades

The textile industries over the course of the eighteenth century developed a great diversity of structures of work organization, with which they incorporated a suitable variety of types of technical change. They form the historical archetype of the transition to factory production and steam-powered mechanical techniques. But this archetype or model of industrialization applied in the main period of the Industrial Revolution to only one part of the textile industry, namely cotton, and even here the transition was complicated and many-sided. There was much forward, backward and sideways motion as techniques were developed and manufacturing organization chosen. Textiles, moreover, provide the archetype for 'proto-industrialization', and in more ways than one. For in the case of cotton, they saw the transition from rural domestic manufacture to the factory system while also encompassing the opposite road of transition into industrial involution or sweating (as in framework knitting) or into outright repastoralization in the West Country cloth industry.

The model of proto-industrialization has always concentrated by and large on textile production. But if we take the other great prototype of early industrial organization – Marx's model of manufactures – the industries appealed to were those containing workshops of skilled artisans, especially in the metal manufactures. As has already been said, Marx drew attention to the engineering workshop of the late eighteenth century and the early nineteenth for all the key features of 'manufactures'. He endorsed Andrew Ure's praise of the 'machine-factory' which displayed the division of labour in manifold gradations – the file, the drill, the lathe having each its different workmen in the order of skill. There was also the special prophetic feature that 'this workshop, the product of the division of labour in manufacture, produced in its turn – machines'.[1] We have argued that in spite of his allusions to rural industry and centralized production, Marx conceived of manufactures in terms of a large workshop in the hands of a capitalist and organized on the basis of wage labour, with the rhythm of the production process still, however, dictated by the craft skill of the workers.

The correspondence between the metal trades and Marx's phase of manufactures was limited in much the same way as was that between textiles and proto-industrialization. But the metal manufactures pose a different range of problems concerning work organization and technological change. In the first place the development of a technology dependent on manual skills was not, in the metal manufactures, in contradiction to the division of labour. For certainly manufacture was 'divided', but this did not necessarily make the labour on component parts any less skilful labour. The skill lay increasingly in the accuracy and precision manufacture of parts of an object which were then 'fitted' with others into the whole. However, the metal manufactures also transcend any simple connection with the large-scale handicraft workshop Marx imagined. Of course, these existed, but the size of units ranged from the outset from the very small domestic or garret manufacture through to the large factory, and so too did their forms of work organization.

The Birmingham hardware trades, Sheffield cutlery, the brass trades and engineering all conjure up images of a workshop culture, a high degree of skill, a large predominance of urban manufacture and the endurance of small-scale production.[2] The metal manufactures pose their own multifarious forms of specialization and development. These industries really were the locus of Nathan Rosenberg's 'continuum of small improvements', or anonymous technical change. They locked into and benefited from that innovative capital goods sector which Rosenberg credits with the saving of capital over the whole economy, and with providing the prime mechanism for technological diffusion. Here we will explore the origins, organization, technologies and labour forces of some of them.

What we will discover in delving into the metal manufactures in this period is the development of technologies based simultaneously on skill and the division of labour, and a combination of artisan independence with the rise of the medium- and large-scale workshop or factory. I will look at the significance of the metal manufactures to Britain's road to industrialization and explore some of the characteristics of skill, work organization and technology in the engineering trades. I will then look at the role of the artisan and the development of various work settings in the trades. I will go on in the following chapter to discuss these issues in depth in the case of some of the Birmingham hardware trades, notably the 'toy' manufactures which established Birmingham's reputation as the 'other face' of Britain's Industrial Revolution.

Metals and manufacture

One of the key technological principles assigned by Landes to the Industrial Revolution was the substitution of mineral for vegetable or animal raw materials, that is, the shift from a wood to a coal fuel technology, and with this the shift from the use of wood to iron and steel in machinery, tools, implements and other manufactures. The development of an iron- and coal- based technology, one of the vital characteristics of the transformation of British industry in the eighteenth century, was particularly noted by foreigners. F. and A. de la Rochefoucault-Liancourt in their *Voyages aux Montagnes* in 1786 admired the English for

> their skill in working iron – the great advantage it gives them as regards the motion, lastingness and accuracy of machinery. All driving wheels and in fact almost all things are made of cast iron, of such a fine and hard quality that when rubbed up it polishes just like steel. There is no doubt but that the working of iron is one of the most essential of trades and the one in which we are most deficient.[3]

With the effective processing of iron and widespread adaptation to its use in Britain went a great facility in the use of tools and mechanical contrivances.

J. R. Harris and Peter Mathias have stressed the empirical characteristics and the adaptive innovative content of eighteenth-century technologies which relied above all on the skilled worker.[4] This was especially so in powered and mechanized processes using coal. The foundation of the Industrial Revolution in coal, iron, steam and machinery relied at its heart on a core of practical skill which was very difficult to transmit outside the mind of the skilled workman himself. Manuals, patents, scientific observation and journalistic description were useless without the skilled worker who actually applied and adapted the technology. It is what P. Courtheoux, writing about iron puddling, has called the separation between knowledge of a process and ignorance of a craft. Methods might be described, but so much of the skilled worker's real knowledge was 'breathed in the atmosphere where he lived' rather than ever consciously formulated. As a French inspector of industry wrote in the 1820s after visiting the Sheffield region:

> These [simple workmen] are truly the metallurgists of Yorkshire and it is among them that one can gather the elements of steel making. But there, as elsewhere, there is barely a common language between the workman and the savant; it is, for example, extremely difficult to determine in many cases what qualities a workman means when he says that iron has 'body', is 'sound', 'strong', 'tough', etc.; all of these, however, are expressions which have a very precise meaning . . .

In the case of the iron puddlers, their skill was essential to the transformation of pig iron into wrought iron. The process consisted of stirring the molten pig iron on the bed of a reverberatory furnace. The puddler turned and stirred the molten mass until through the decarburization of air, circulating through the furnace, it became converted into malleable iron. Puddlers, underhands, rollers, and all the other iron tradesmen were tough, rough and uneducated.

But if uneducated, [they] were not ignorant. . . . A new man in the trade started to learn in earnest, the hard way, by doing, not talking, and he developed a taciturnity which lasted all his life.[5]

Richard Cobden in the nineteenth century hailed the metal industries and the skilled workers of the eighteenth:

Our strength, wealth and commerce grew out of the skilled labour of the men working in metals. They are at the foundation of our manufacturing greatness.[6]

These special skills applied not just to the processing of metals, but also to working in them. It is to the latter that we shall largely refer. For the large scale of the processing works, their use of mechanical and powered techniques, their centralization of processes, and the capitalist organization of their workforces form a separate problem and story in itself. It is, furthermore, one which has often been told in the standard histories of the Industrial Revolution. For metal processing or heavy industry forms the other side of the coin of factory production. But the metal-processing works also spawned a whole series of relatively small-scale metalworking industries using iron, steel, brass, copper and various alloys. These relied on skilled workers, diversity of output, innovative practices, and handtools, and were organized within artisan and putting-out systems. Many of these industries remained organized on a workshop or family production principle throughout the Industrial Revolution, simply adapting steam power and factory premises to the dictates of small-scale production. Metalworking spawned a distinctive form of specialization and artisan skill within elaborate networks of small-scale production side by side with large-scale works.

Engineering
An early close relationship grew up between the heavy industries and the infant engineering industry. The first great ironworks – Coalbrookdale, Carron and Bersham – were also the first

builders and users of machine tools.[7] And the pinnacle of the skill and independence of the worker in metals was the early engineer, especially his forebear, the millwright. William Fairbairn, one such millwright, argued that in England before the eighteenth century the most important articles of machinery such as windmills and watermills were brought over from the continent. And as such contrivances became more used, a special class of native artificers sprang up to attend them. These were called millwrights, and they designed and erected windmills, watermills, pumping apparatus and various kinds of rough machinery. They were the first who devoted themselves exclusively to engineering work.[8] Where the staple material of the millwright had once been wood, after the mid-eighteenth century he adapted to the increasing use of iron in machinery.

The origins of engineering in the millwright's craft were often restated. A journeyman smith reporting to the Committee on Apprentices in 1813 said he was called a machinist and engineer, that the business was a new one, that the millwrights had set up in it, employed smiths and made steam engines, lathes etc.[9] As Jennifer Tann has written, the millwright of the early Industrial Revolution worked in wood. He was concerned with the application of power to an industrial process and the transmission of that power to machinery, in other words, the linking of the prime mover via a system of shafts and gears to machines. The millwright was frequently called on to assess the power requirements of particular machines, and not infrequently the plan of a mill and its layout of machinery.[10] And Fairbairn described the millwright of former days as the 'sole representative of mechanical art', 'a kind of jack of all trades, who could with equal facility work at the lathe, the anvil or the carpenter's bench'.

Thus the millwright of the last century was an itinerant engineer and mechanic of high reputation. He could handle the axe, the hammer, and the plane with equal skill and precision; he could turn, bore, or forge with the ease and despatch of one brought up to these trades, as he could set out and cut in the furrows of a millstone with an accuracy equal or superior to that of a miller himself.[11]

The millwrights who turned their hands to the mechanical and power needs of the woollen and cotton industry soon turned to the manufacture of machinery itself. The progress of the eighteenth-century woollen industry, premised on increasing quantities of a limited range of machinery, brought some country millwrights into the manufacture of textile machinery on a small scale. And in the early stages of the transfer of the cotton manufacture to the factory system, the traditional country millwright, such as Thomas Lowe, was used. The types of millwrighting needs thrown up by the various branches of the textile industry in turn affected the subsequent regional development of the engineering industry. The demand for millwrights was greater in Lancashire than elsewhere. In Yorkshire the smaller, less specialized millwright continued to cater to country mills, and in the West of England the limitations of the country millwright and small machine maker, who formerly satisfied most needs, were made clear only when the cloth manufacturers began to investigate possibilities of adopting steam power.[12]

Where the early practice of the machine maker had involved carrying his tools, including lathe, drill and grindstone from mill to mill in order to build the machinery on the spot, now textile machine shops grew up around the textile areas. This led to a new differentiation in workers' trades. Pattern makers, iron and brass founders, smiths, hammermen and firemen, vicemen and filers, turners and planers came to take the place of millwrights, carpenters and blacksmiths.[13]

In fact, many of the first engineering workers owed their origins not to the millwrighting, but to the smithing trades. The smith's trade was already subdivided by the eighteenth century into the forgework of the fireman, the filing and finishing work of the viceman, and the work of the hammerman. It was the viceman who was the ancestor of the fitter, and the largest department of most nineteenth-century engineering shops remained the smithy.[14]

Alongside the smithing trades, there were metal trades long established in making tools, found in southwest Lancashire by the seventeenth century. Peasant toolmakers to the west and south of St Helens produced chisels, pliers, vices, gauges, small lathes and files to be used by clock- and watchmakers, whitesmiths, and later machine makers and cotton spinners. The area was also well known for its other rural metal trades, the manufacture of watch parts, locks and hinges, nails, pins and wire. A long-established force of skilled craftsmen trained to work in metal to a high degree of accuracy formed an important start for the machine-making and engine-building industry which soon arrived to service the textile mills.[15]

If millwrights formed one primary source of skilled engineering labour and smithing another, these were not sufficient, for the specialization and training up of workmen was also a major source of labour. This was so, in particular, in the manufacture of steam engines. Early steam engines were erected on the site they were to be used on, and this involved a great deal of fitting and local manufacture of simpler components. Boulton and Watt only supplied a man to supervise the erection of the engine. The first engine shop at Soho was on a very small scale, with only two smiths' hearths, a fitting bench and a single lathe. But the new Soho Foundry of 1795 was the first heavy-engineering shop; it concentrated on the design of patterns for castings and the manufacture of parts calling for accuracy in machining and skill in fitting, such as valves, valve chests and parts of the valve gear.[16] Watt constantly complained about the clumsiness and incompetence of his workmen, and so systematically set about training up a body of them. He tried to keep certain groups to special classes of work and encouraged them to bring up their sons to it.[17] Fairbairn reported that at Soho it was not unusual for some precise line of work to be followed by members of the same family for three generations. This, along with the improvement in tools, meant that

the facilities thus afforded led to constant progressive improvement in the character of the work done, at the same time constantly diminishing the dependence on mere manual skill.[18]

Some of these Soho workmen eventually made their way to Manchester where they took employment as enginemen, thus providing a pool of skilled labour.[19] Later engineering workshops such as that of Maudslay in London, on a small scale in 1798 and a larger scale in 1810, formed a nursery of engineering talent for the nineteenth century.[20]

The creation of this skilled engineering labour force involved, therefore, the division of labour and specialization of function. Nor was this a recent development. The early metal trades of the St Helens region and Warrington were already highly subdivided by the seventeenth and eighteenth centuries. The manufacture of watch movements and tools was 'put out' as early as the seventeenth century to rural workers in southwest Lancashire by all the big watch firms in London, Coventry and Liverpool. The files produced in Warrington in the eighteenth century were sometimes made from start to finish by the same outworkers, but more often some workers confined themselves to forging and others to cutting, and then sent the files elsewhere to be hardened.[21] By the mid-eighteenth century most British watchmakers ordered the rough movements of the watch from specialist firms or outworkers in Lancashire. These worked from paper measurements using their own Lancashire gauge. 'Pinion wire was made at the mill; springs, by artisans who did nothing else; chains, by chainmakers (often women); cases by casemakers.'[22] In eighteenth-century engineering there was the division between fitters and turners. The fitter was acknowledged as a highly skilled craftsman, but at Soho fitting was subdivided in such a way that to each group of fitters only one article or group of articles was assigned. Similarly, turners were long known as general workmen of whom a high degree of skill was expected, and they did most of their boring on ordinary lathes. But at Soho a number of different classes of boring were separated out. By 1824, Galloway, a London engineer, had between 75 and 115 employees, and these were all divided into a number of different trades – pattern makers, iron and brass founders, smiths, firemen, hammermen, vicemen, filers, brass, iron and wood turners. There was, in addition, in areas like Manchester, a further division of the engineering trades based on product. Rollermaking and spindlemaking, for example, became separate industries.[23]

The point of much of this specialization was greater accuracy and precision, qualities which became more and more valuable with mechanization. It was a 'transition from the workmanship of chance to the workmanship of precision'. 'The craftsman in gold and silver could compensate for differences in the quality of his metals: the constant motion of a machine required constant materials.' But the transition was a slow one, for when William Fairbairn went to Manchester in 1814, 'the whole of the machinery was executed by hand. There were neither planing, slotting nor shaping machines; and with the exception of very imperfect lathes and a few drills, the preparatory operations of construction were effected entirely by the hands of the workmen.'[24] Chipping and filing were the major craft processes behind much of engineering, and with this the chisel and file comprised the major part of the workmen's 'fixed capital'. Until Maudslay perfected a screw-cutting lathe at the beginning of the nineteenth century,

> the tools used for making screws were of the most rude and inexact kind. The screws were for the most part cut by hand: the small by filing, the larger by chipping and filing . . . and each manufacturing establishment made them after their own fashion. There was an utter want of uniformity.

These tools were owned by the workmen, 'the men were Masters', and there were large numbers of them. In 1825 there were 400 to 500 engineering masters in the London area, employing no more than 10,000 men.[25]

Hardware and cutlery

The continued importance of skill in spite of specialization in the engineering trades was echoed in the Black Country and

Birmingham hardware trades and in the Sheffield cutlery trades. Regional differentiation in terms of levels of skill appeared very early. By the end of the seventeenth century, Birmingham was tending to produce articles requiring a great deal of skilled labour, but few or low-cost raw materials and little transport, while cruder manufactures migrated to the rest of south Staffordshire. The cutlery trades too were divided between the high-class trade of Sheffield and the inferior common trades of the countryside.

Skill became differentiated, but the characteristic unit of production remained relatively small in scale.

> The variety in the design of the metal articles is so great, the opportunities for standardized production so few, and the importance of skill so evident, that the small unit is naturally to be preferred.

In gunmaking, for instance, an extensive division of labour was elaborated within a structure of small-scale units of production. The gunmaker owned a warehouse in the gun quarter, acquired semi-finished parts and gave these out to specialized craftsmen who undertook the assembly and finishing of the gun. These in turn bought their parts from a whole range of independent manufacturers – barrelmakers, lockmakers, sight stampers, triggermakers, ramrod forgers, gun furniture makers and bayonet forgers. The combination of small-scale production with a high degree of fairly flexible skill meant both that manufacturers could throw the burden of depression on to the workers themselves, and that workers could turn their skills in the use of the file and the lathe to some other trade.[26]

Skill, in a similar fashion, determined the location not just of the various hardware trades, but also that of the brass and copper trades. These skills were, however, evidently spread more widely, for by the early eighteenth century many towns carried on the manufacture of brass, with no other special advantage than a resident class of artisans already skilled in working in metals. There were by that time large numbers of braziers and coppersmiths throughout the country, since more brass and copper was produced than could be economically consumed by the integrated brass founding and working establishments.[27]

Many of the metal trades, like textiles, were carried on in combination with some form of pastoral agriculture. A division of labour between the agricultural and industrial pursuits of workers developed more or less rapidly, depending on the fertility of the soil. Metalworkers in the Birmingham area kept less capital invested in agriculture than did Sheffield craftsmen in the better farmland of south Yorkshire and north Derbyshire.[28] But even these trades became more and more divorced from agriculture over the course of the seventeenth to the late eighteenth century, and craftsmen became increasingly dependent on the cash incomes derived from their forges.[29] By 1801, for instance, the towns and villages of eleven Black Country parishes contained 80,000 people, or one third of Staffordshire's population. The area contained very large numbers of nailers (40,000 were employed in the trade in the Midlands by 1800), as well as a great range of trades from bucklemakers, chainmakers, buttonmakers, tinners and platers, japanners and pinmakers. The population of these industrial villages was by this time 'anonymous'; it was also 'very mobile, as the rapid expansion of the area attracted much casual labour, and technical change attracted populations with particular skills'.[30]

Just as the division of labour proceeded in engineering along the lines of process and product, so too in the cutlery manufacture which developed rapidly in Sheffield and its hinterland from the seventeenth century onwards. In that century forging, grinding and halfting had all been kept in the hands of the master workman. But by its latter half, the industry had divided into three trades: those producing knives were separated from the scissormakers, and these in turn from those making shears and sickles. Grinding was separated out very early in the scythe trade, that is, by 1630–50, but in most of the other trades this did not occur until the mid-eighteenth century.[31]

The skills of forging and grinding, which were gradually separated out for many of the cutlery trades from the latter half

Sheffield cutler finishing a knife
(George Walker, *The Costume of Yorkshire*, 1814)

of the eighteenth century, were also generally divided into different premises. The first process, forging, was performed, depending on the product, by one or two workers. Heavier articles, such as table blades, files, tools and larger instruments, required the assistance of a striker or butty who manipulated a heavy double-headed hammer. The second process, grinding, was performed in separate departments of larger works or in separate establishments called 'wheels'. In earlier 'wheels' handpower was frequently used, with a boy turning a large fly wheel, or the 'wheels' were scattered along the banks of streams, using water power. The first steam 'wheel' was set up in 1786 and soon yielded up cases of grinders' disease, for unlike the old methods of powering the grindstone, steam power compelled uninterrupted work day after day. This was reinforced by the continued use of drystone grinding for the lighter implements until 1840.[32]

Grinding was followed by further refinements of the blade with hammer, whitening stone, emery and buffing wheels. The final process of building up or halfting the knife was the true cutler's art; it relied only on handskill and basic tools – drills for boring, files, vices, a glazer and buffs.

The building up of a pocket knife is a task of greater complexity and one for which it is not easy to substitute mechanical process owing to the infinite variety in which these goods must be produced in order to meet the whim of the purchaser: a single firm may supply such knives in thousands of different patterns.

File cutting was a particularly skilled process, yet divided between forger, grinder and cutter. The forger moulded the steel into the right shape, the grinder ground it on a wet stone, and the cutter cut the teeth into it with hammer and chisel.

By a succession of smart blows parallel ridges of astonishing conformity and exactness are produced, the blows following one another with wonderful rapidity.[33]

This retension of high degrees of skill in spite of the division of labour between process and product entailed conditions of semi-independence among artisans and complex systems of industrial organization in the metal trades. But before the advance of steam power and the special dependencies created within piecework systems, the proverbial independence of the skilled metalworker was by the nineteenth century becoming more and more illusory.

Artisan independence and skill

The independence of the early millwrights was proverbial; it went with an autodidact tradition and strictly enforced trade-society rules. Fairbairn reported that the millwrights of the turn of the nineteenth century formed their own 'Millwrights' Institutes' in each shop.

> On the more peaceful occasions, however, it was curious to trace the influence of these discussions on the young aspirants around, and the interest excited by the illustrations and chalk diagrams by which each side supported their arguments, covering the tables and floors of the room in which they were assembled.[34]

They formed their own benefit and trade societies which dictated the hours of work – light till dark in winter, and six till six in summer – as well as the rate of wages, with no member allowed to work for less than 7s a day.[35] Watt bemoaned the millwrights' solidarity when they struck work in 1795:

> At some places they have left work, tho' the proprietors would give them their own terms, but they will not work till every master accedes to their proposals; they are already better paid than any other class of workmen, having a guinea per week and an allowance of 6d per day for beer, and they always make 7 days a week.[36]

The trade in some places was virtually a patrimony, and Fairbairn found it extremely difficult to gain entry himself to any of the trade societies in London which between them controlled admission to employment. But once entry was gained, the way was open to independence and eventually to individual proprietorship for those who could amass a small capital. After employment as a journeyman for several years in Manchester, Fairbairn in 1817 set up on his own with James Lillie. Taking on a number of small jobs sufficed to enable them to make a lathe capable of turning shafts three to six inches in diameter, and by

1824 they had acquired a sixteen-horsepower steam engine.[37] Maudslay's rise to the proprietorship of a London engineering workshop was similar. He worked nine years for Bramah and first set up on his own in 1798 to make ships' blocks. Over the next ten years he gradually branched out to the manufacture of calico-printing machinery, small engines, and power-driven lathes. He went into partnership with Joshua Field, making improvements in the lathe and expanding his business into marine engineering. His works soon acquired a reputation for the skills of the engineers trained there; they turned out Richard Roberts, David Napier and Joseph Whitworth among others.[38]

The strict entry qualifications of the millwright and engineering trades contrasted sharply with the general lack of formal restriction in the hardware and cutlery trades, yet a form of independence still prevailed among artisans. In cutlery, the rapid expansion of the industry in the early eighteenth century undermined the power of the Cutlers' Company to impose restrictions. An attempt was made in 1791 to incorporate all those currently practising the trade into the Company, and to impose strict entry qualifications subsequently. But this, too, broke down, and in 1814 apprenticeship restrictions were abolished. Apprenticeship regulations were similarly difficult to impose in many of the hardware trades. There were guilds among the metalworkers in some parts of the West Midlands. But the wide scattering of raw materials, the simplicity of early processes, and the part-time characteristics of the occupation made any attempt on the part of the towns to monopolize the industries very difficult. The absence of formal apprenticeship restriction did not, however, imply the sway of free market forces, for it was here that workshop custom imposed its own structures. The notorious 'rattenings' of the Sheffield trade unionists, that is, the removal of the tools and wheel bands of recalcitrant members of the trade, had its origins in the legal right of the guilds to enforce their rules by the removal of the property of offenders.[39]

This artisan independence in the metal trades was reflected in the forms of industrial organization which developed through the eighteenth and part of the nineteenth centuries. The line dividing outworker and pieceworker, manufacturer and artisan was ill-defined. Clapham long ago pointed out the extent of semi-independence in the metal trades. The journeyman in the Sheffield trades might be an outworker or a pieceworker, but he had his own tools and forge. The small master in the Birmingham trades was independent though he might be working pretty regularly for a particular factor, and masters worked regularly on the employers' premises, as at Crawley's works in Winlaton, County Durham.

> Masters got tools and materials from a works iron keeper, then employed their own hammermen and prentices, and were credited with the selling price of their goods less cost of material and some overhead to Crawley.[40]

Customary trade practice particularly over apprenticeship regulation was vital to artisan independence. Friendly and trade societies enforced customary demarcation lines on wages, prices and employment policy, exerting some degree of workplace control. And in addition, 'independence' was written into the structure of work organization, so that every skilled artisan was conceived of as a small master.[41]

The conditions of this semi-independence, however, varied with economic fluctuations, and became more constrained with the development and increasing control of intermediaries. A vital aspect of this control came through indebtedness. In the words of one historian, 'Putting out was necessarily a form of credit, though historians have often treated it loosely as a kind of wage labour.'[42] Outworkers who had the nominal independence to seek out alternative suppliers of raw materials were frequently bound in debt to one employer. Among regular workers as well as more casual workers, debt to an employer or supplier was not temporary but long-lasting. Stubs, the Warrington filemaker, spoke of one workman tied by debt to another firm of toolmakers, and so anxious was he to work for Stubs that he was willing to transfer to him his rights to the labour of his sons who were apprenticed to their father. In the pin trade, parents borrowed on the security of their young children. As T. S. Ashton put it:

In other industries payment in truck or the new discipline must be given first place among the ills afflicting the wage earner, in the metal working trades indebtedness to the employer would seem to have been by far the most serious barrier to the attainment of economic liberty.

Courts existed for the recovery of small debts. Stubs used the Warrington Court Baron for the Recovery of Small Debts to obtain repayment of loans, and in addition for redress against workmen returning poor or insufficient work. The Birmingham small debts court, the Court of Requests, met from 1752 and dealt with 80 to 100 cases a week.[43]

Neither was the existence of a proliferation of small men a sign of 'independence'. The 'little masters' of Sheffield, for instance, multiplied not in times of prosperity, but in those of commercial stagnation and distress. These little masters might be merely factors using the labour of outworkers, or they might employ a small team and rent a room in a factory.[44] And though the Sheffield artisan might define his independence by his ownership of his own tools, it was also increasingly apparent by the nineteenth century that a local group of merchant capitalists controlled the circulating capital of the trade and the distribution of the finished product, and that furthermore industrial sites were in the hands of various rentier groups.[45] In the hardware trades the 'independence' of the artisan lay open to the abuse of the discount system, whereby manufacturers demanded a discount on goods 'sold' them by their artisans – a form of wage-cut. The small master was, furthermore, placed at a greater disadvantage in buying raw materials and selling his product. He increasingly became an outworker, to whom the larger manufacturer had no commitment in terms of capital or marketing networks.[46]

The reliance of the metal trades on the skill and semi-independent status of their workforce entailed complex systems of industrial organization based on the small-scale unit of production and interlocking networks of intermediaries. Capitalist expansion and industrialization in the metal trades found their context not in the factory but in the garret master and various forms of sweating. On the one hand the system of outwork prevailed because of its superior advantage for the production of specialities on a small scale.[47] On the other hand, this advantage was one which could allow the garret master and the 'slaughterhouse men' to undermine the substantial workshop. In lockmaking, for instance, most employers were small masters with one to four apprentices and one or no journeyman. When apprentices had served their time, they generally had no alternative but to set up as small masters. There were many small masters and many apprentices in the trade. This was not just a feature of lockmaking, for in the latter half of the eighteenth century the custom in the metal trades in most parts of England had grown up for workmen as well as employers to take apprentices.[48] In the Birmingham trades the small garret master became more significant in the early nineteenth century, whereas earlier a large number of substantial workshops had thrived. This point, which is discussed at some length in the following chapter, is missed by most historians who assume an enduring small-scale character for the Birmingham manu-factures throughout the eighteenth and nineteenth centuries.[49] In fact, industrialization brought a form of dualism whereby both large-scale and extremely small-scale firms probably grew at the expense of substantial artisans and medium-sized manu-facturers. The appearance of more and more small garret masters was encouraged by the proliferation of various inter-mediaries and agents; it took its worst form in nailmaking where the 'fogging system' came into prevalence in the nineteenth century. The role of the factor ranged from the one concerned only with the distribution of the product and with financing the little masters, to the type who exercised an intimate control over raw materials and the coordination of labour. His major role was in finance where he formed virtually the only link between the small masters and the banks.[50]

The independence of small masters was in many cases mythical by the nineteenth century. Their working days and practices were constrained at every point by the factors, whose power was now if anything increased by the availability of steam power. For steam power in the metal trades could be hired out with room or bench space, and the worker now had to come to the place where the steam engine was housed and work the hours that the steam engine was run. Just as his markets, his raw materials and his capital were out of his own control, so too now were his place and his hours of work.[51]

If anything, factory production in the metal trades entailed just as much if not more of the traditions of independence of the domestic worker. In many cases the manufacturer grew out of a factor and remained at heart in this capacity. He did not become deeply involved in the details of manufacturing processes, and his employees carried on the traditions of domestic manufacture by providing their own tools and paying for their workspace. Deductions were frequently charged in factories against the wage for shop room, gas and power, and in foundries head casters customarily paid for the use of the sand mill.[52] Indeed, subcontracting was very common in the centralized units of the heavier metal industries, just as it was in the decentralized light industries. In the brass manufacture, the head caster paid and supervised his own moulders and labourers; journeymen were employed on a payment by results system, and underhands were employed by journeymen at daywork rates. In a brass-finishing shop of a larger works, the journeyman was expert in his branch of the trade. He organized the work of the shop and got it done efficiently; forged, ground and hardened his own tools and kept his lathe or spindles in good order. He was designer, supervisor, toolmaker, tool setter and all-round workman.[53] In the centralized units of the light industries, too, piecework prevailed. Women piecemasters ran groups of workers in buttonmaking, papiermâché and lacquering shops of larger 'toy' factories.

In the famous engineering works of the day much the same combination of artisan independence and factory production prevailed. Eric Roll described the shop structure at Soho as a transition between handicraft production and modern mass production, for only one basic operation was performed within each shop, and though they were located according to a systematic order, they were still semi-independent units. This was combined with elaborate systems of piece wages. At the Soho

Foundry men were engaged at a certain weekly wage. Another agreement was then made with the foreman to perform a job at a set rate or faster.[54] Maudslay's shop was actually organized by Samuel Bentham, a specialist in production organization who is often credited with the origins of the assembly line. Maudslay's example was closely followed by his student James Nasmyth in the nineteenth century, who demanded that his building be built 'all in a line'. 'In this way we will be able to keep in good order.'[55]

Factory production, the proliferation of tools and the application of steam power were regarded by some overoptimistic industrial pundits of the day as the means of overcoming the special needs for skill and independent status of workers in the metal trades. Fairbairn spoke for these when he declared:

> The improvements in tools changed the mode of doing mechanical work, by rendering necessary large and carefully laid out manufactories. The old millwright had little need of large or expensive premises or plant... but under the improved conditions brought about by Watt's inventions ... it was necessary to have more systematic arrangements ... and these necessities brought about the establishment of large manufactures which gradually supplanted the old millwright's trade.
>
> In these manufactures the designing and direction of the work passed from the hands of the workman into those of the master and his office assistants. This led also to a division of labour, men of general knowledge were only exceptionally required as foremen or outdoor superintendants; and the artificers became in process of time, little more than attendants on the machines.[56]

Eric Roll, however, went out of his way to emphasize the similarities between the production arrangements at Soho and some of the principles of scientific management of the late nineteenth and early twentieth centuries; nevertheless, he still disputed the decline of skill.

One may therefore be led to question the common assertion that the skilled worker is gradually dying out, an assertion usually accepted without any convincing evidence. If the elimination of the skilled worker is taken as one of the results of the use of machinery, then it must be evident that he has taken a very long time in dying, since the tendency for his skill to diminish has been in operation very much longer than is generally assumed.[57]

And as David Landes has pointed out much more recently, even after the invention of power tools 'each craftsman remained judge of his own performance', his specifications were approximate, and he filed down each part to fit the whole.[58]

Notwithstanding these complicated patterns of artisan independence, it is true that important and far-reaching changes took place in both work organization and in technology in the engineering trades and the hardware trades in the 1820s to 1840s. In engineering, there was a drastic reduction in the number of engineering employers after 1825, and the diffusion of machine tools from 1830 led to larger and more heavily capitalized engineering firms. Along with this, the main locus of the industry shifted from London to south Lancashire. 'In the period 1830–50 the British engineering industry was transformed from a labour- to a capital-intensive one.'[59] There were similar developments in the Birmingham hardware trades, where in the first half of the nineteenth century larger establishments were introduced into most of the town's staple trades. As their activities expanded, this served to increase the tempo of the competitive relationship at all levels of industry. Along with such large firms there was a proliferation of small garret masters and sweated labour, with 'the ranks of the little makers swollen periodically with the unemployed attempting to avoid the parish by becoming garret masters'.

The metal trades may well have been a stronghold of handicraft skill and artisanship, but these ideals of the 'manufacturing mode of production' were never static.[60] They underwent enormous change in the early nineteenth century, but likewise they did so too in the middle of the eighteenth. The industrialization of the metal trades took on its own special form within the framework of artisanship and handicraft. But even before industrialization these two qualities were in a state of flux. Nothing better epitomizes this 'movement' of transitional manufacturing processes than do the Birmingham hardware trades. We will now turn to an important part of them – the 'toy' trades.

CHAPTER 12

The Birmingham Toy Trades

The toy trades cover a range of new products which became identified with the Birmingham metal industries in the eighteenth century. It is difficult to give a precise definition of them. The *Dictionary of Arts and Sciences* in 1754 defined Birmingham wares as 'all sorts of tools, smaller utensils, toys, buckles, buttons in iron, steel, brass etc.'. John Taylor and Samuel Garbett told the House of Commons Committee of 1759 that toys were articles in which gold and silver were manufactured, but stressed that the gold and silver used accounted for only 1–5 per cent of the value of the product. A more recent definition of the eighteenth-century toy describes it as a comprehensive term

for an assemblage of numerous kinds of more or less useful wares, of small dimensions, and varying from a few pence to many guineas in value. It included much of what is now termed jewelry, small articles of plate, sword hilts, guns, pistols, and dagger furniture, buttons, buckles, bracelets, rings, necklaces, seals, chains, chatelains, charms, mounts of various kinds, étuis, snuff boxes, and patch boxes.

It was also considered to cover any article with the common characteristic that its patterns changed with great frequency. *The Victoria History of the Counties of England* breaks the toy trades down into heavy (wares of polished iron) and lighter (steel). The heavy trades produced the 'tools and instruments used by carpenters, coopers, gardeners, butchers, glaziers, upholsterers, farriers, masons, plumbers, coachmakers, millwrights, saddlers, harnessmakers, tinmen, shoemakers, weavers, wire drawers and wheelers'. The light steel toy trades, said to have occupied half of Birmingham before the French Revolution, produced 'buckles, purse mounts, brooches, bracelets, watch chains, key rings and swivels'. The toy departments peculiar to Matthew Boulton's large hardware factory included clocks with ornamental cases, fine art bronzes, filigree work, buttons, buckles and clasps, lamps, bracelets, candlesticks, and tea irons.[1]

The toy trades were a new industry, based on international and especially the new colonial markets, making Birmingham in some senses 'proto-industrial'. But much more striking is the way in which these new industries were based on a special form of technical change integrated with a complicated structure of small-, medium- and large-scale enterprises. Their rise formed a part of the wider development of the metal industries on the one hand, and of Birmingham on the other in the eighteenth century. Aside from the rising domestic demand for all kinds of metalwares, there was also a rapid rise in exports. Metal goods made up 7 per cent of manufactured goods exported in 1722–4, and this had increased to 14 per cent by 1772–4. The Midlands was already exporting large quantities of fashion goods early in the century, for there is evidence from 1712 of exports of watches, locks, clocks, buckles, buttons, and other brass toys from the region to France.[2] Birmingham's place in the development of these industries is variously attributed to its geography, its unincorporated political status, and its role as a haven from religious persecution. Transport problems put a premium on articles where skilled labour was an important element, and this is supposed partly to have accounted for the predominance of the manufacture of small articles showing the hand of the skilled worker. The lack of incorporation of this relatively new town attracted both labour and capital which faced barriers to entry and other restrictions in many areas. Birmingham's early reputation for religious toleration formed yet another attraction for them. It was for this reason that the buckle industry was introduced from Staffordshire, for craftsmen driven from Walsall by religious persecution started making buckles in brass and copper in Birmingham.[3] The decline of the buckle manufacture owing to changes in fashion in the 1790s made room for a new expansion of the button manufacture which took over the same skills, labour force and raw materials. By 1759 there were about 20,000 employed in the toy trades in Birmingham and the surrounding district, and the value of the ornamental part of the trade was estimated at £600,000 per year, £500,000 of the total value being derived from exports.[4]

In addition to the widespread production of buttons and buckles there appeared other new toy branches distinctive to Birmingham. Japanning, a form of enamelling with a material made of a mixture of turpentine, balsam, oils, pitch, resin and wax, was started in Monmouthshire and spread to Staffordshire in the early eighteenth century. Initially executed in tin plate by hand, it was only with the introduction of papiermâché that it became a cheap, mechanical and profitable trade.[5] The stamped brass foundry trade, using sheet metal, die and hammer, was introduced by the London toymaker John Pickering and spread so rapidly in Birmingham that by 1770 the town housed stampers, die sinkers, stamp and press makers, and button stampers. Plating silver over copper also became popular in the button and buckle trades, the manufacture of snuff boxes and other products.[6]

The organization of the toy trades

To what extent were the toy trades a special case whose organization, technology and labour market were so different from the norm that their divergence from the model patterns of pre-industrial change only serves to prove the rule? There was certainly enough variety in the organization of these trades for them to fit the criteria of both the Marxian model of manufacture and the proto-industrial model. On the one hand, they were in some cases organized in the ideal large-scale workshop and 'factory', displaying all the subdivisions of process and division of labour which Marx described. On the other hand they might equally be found in small household workshops, combining a long-standing connection with the land, with a new dependence on large and sophisticated mercantile networks, thereby yielding an admirable example of the other model.

Marx's allusion to the engineering workshops as the epitome of 'manufactures' might just as well have been to the toy factories of John Taylor and Matthew Boulton. Taylor reported to the House of Commons Committee in 1759 that he employed 600.[7] Though some of these may have been outworkers, his factory was large and it was not alone, for Boulton by 1770 employed 800 to 1,000 at Soho. The Soho workshops were described in the *Directory* of 1774 as consisting of 'four squares with shops, warehouses etc. for 1,000 workmen in many different branches'. Elliott and Sons of Frederick Street were said to have employed several hundreds, mainly women, on three floors.[8] Though most manufacturers in the town were relatively small men with a capital of less than £100, by the last half of the eighteenth century there were a number with great wealth. By 1783 ninety-four manufacturers in the town had a capital greater than £5,000, eighty more than £10,000, and seventeen £20,000.[9] The large scale of enterprises in these new trades derived encouragement notably from the variety and expense of raw materials, and from the specialization of artists and finishers. Many of the processes were not suited to the family enterprise, and expensive skilled labour was most efficiently used where there was some degree of division of labour.[10]

Birmingham was a locus of urban artisan manufactures, but it was also in many ways proto-industrial. In the first place, its manufactures were exported in the main outside the region, in particular to the new American markets. And like the textile towns it interacted closely with a rural industrial hinterland. There was an easy transition in the Midlands from the lockmaking or bucklemaking trades to toymaking. Families that first appeared as boxmakers, japanners and toymakers, both in the countryside and in Birmingham, were almost all sons of bucklemaking, lockmaking and yeoman families in Sidgley, Bilston and Wolverhampton. Most of these families had held customary or freehold land locally, and could raise small mortgages on it for capital. Immigrants from the counties of Warwickshire, Staffordshire and Worcestershire accounted for 500 out of 700 settlement certificates from 1686 to 1726. The town was also connected with a wider hinterland. For the rapid growth of the copper and brass industries in the Bristol area, Warrington in Cheshire, and Cheadle in north Staffordshire was closely linked to the rise of the Midland toy trades.[11]

The large size and extensive division of labour in some of the toymaking firms very likely also derived from older traditions in the integrated foundry and metalworking firms. Crawley's ironworks, which employed very large numbers of ironworkers, nailmakers and other hardware manufacturers, was a case in point. The rolling and slitting processes of ironworks were for the most part separated out early from ironmongery. The rolling and slitting mills were highly centralized and capitalist firms. Ironmongery and particularly nailmaking were organized within a domestic system; but coexisting with this were centralized works of various sizes. Work was organized in the nailmaking industry, on the one hand, through a direct merchant-employer relationship, and on the other through an indirect system using agents.[12]

The brass manufacture, which was very important to the emergence of the Birmingham toy trades, had a similar heritage of large-scale integrated works. The two stages of the brass and copper industries – first, the mining and smelting of copper and making of brass, and second the working up of copper and brass into finished articles – were organized before and during the eighteenth century in large monopolistic firms such as the Mines Royal. Though brassworking and coppersmithing were largely organized on a domestic basis, they were controlled through company agents and factors. One such company, the Warmley Company, in 1767 employed 800 at one site in Warmley and another 2,000 outworkers making copper and brass spelter and utensils in copper and brass. The Anglesey Company employed 1,200 miners in Anglesey, it had smelting works at Amluck, St Helens and Swansea, and rolling mills and manufacturing works at Greenfield and Great Marlow.

Brass founding was established in Birmingham in the early eighteenth century, and by 1797 there were seventy-one brass founders in the city. By the end of the century all branches of the brass and copper industry were to be found there. Extensive specialization set in from 1770 according to groups of articles, that is, brass fittings for furniture, fittings for carriages and harness, and fittings for the engineering and plumbing trades, including cockfounding for steam engines. The expansion of the trade had, however, been hindered by restrictive practices on the part of the suppliers of the metal. In the early eighteenth century a combination of brass and copper smelters kept only one warehouse in Birmingham and sold the metal at high prices. By 1780 the brass and copper manufactures in Birmingham formed a substantial sector of medium- and small-scale artisan firms. And while none of these on their own could challenge the brass founders' monopoly, they discovered that together they could by-pass it. Like the scribbling and fulling mills of the West Riding woollen industry, such a brass foundry was necessarily centralized and highly capital-intensive. And the Birmingham manufacturers like the small clothiers discovered that they could have access to the control of such a capital-intensive stage of their manufacture by clubbing together to form their own cooperative concern.[13]

The evidence for large-scale workshops and factories is balanced by other evidence that most firms were very small in scale – part of a workshop-dominated economy. Large-scale processing units might be integrated into this workshop economy via the kind of cooperative organization discovered by the brass manufacturers. It is the small scale of enterprise which supports the theory of upward social mobility from artisan to small master, and which accounts for the community of interest between workmen and employers generally attributed to eighteenth- and nineteenth-century Birmingham.[14] In fact, what seems to have been most common in the Birmingham toy trades was the medium-sized firm. In the first place, handtools could constitute a fairly substantial investment. The number of tools left by bankrupt or deceased firms as advertised by *Aris's Gazette* in the 1780s indicates firms of an order greater than the family firm. William Orchard, a buttonmaker, in January 1789 advertised the sale of his tools, including twenty-one lathes. John Simmonds, another buttonmaker, in April 1769 disposed of three stamps, a press, fifteen lathes, two pairs of shop bellows and various vices and dies. And in June 1789 a bucklemaker had twenty buckle vices, lathes and presses to sell. Those advertising for partners in the button, buckle and japanning trades in the 1780s were seeking partners with a capital to advance in the range of £400 to £600. Among the better known of the medium-sized firms who reported to the Committee of 1812 were those of George Room and Joseph Webster, manufacturers of japanned goods. The one employed 40, the other 100, out of a total of 600 to 1,000 in the trade. Benjamin Cook, a jewellery and toy manufacturer, employed 40 to 50; he claimed there were 7,000 in the trade distributed between 150 masters. Thomas Osler, a glass toy and button maker, employed 80 to 100; William Bannister, a plater, employed 120; and Thomas Clarke, who produced webbing, brass and toys, employed 150.[15]

This range in the size of firms, with a special significance for the medium-sized or fairly substantial business, corroborates the quite extensive division of labour to be found in the toy trades, which appear often to be a very good example of that division of labour, both technical and social. Lord Shelbourne's description of 1766 bears repeating:

> There a button passes through fifty hands, and each hand passes perhaps a thousand in a day – likewise, by this means, the work becomes so simple that five times in six, children of 6 or 8 years old do it as well as men, and earn from 10 pence to 8 shillings a week.[16]

Elliott and Sons were supposed to have carried on fifteen different processes by handpress, including cutting, raising, perforating the metal and placing the paper. Messrs Borwell of James Street, who made gold and silver filigree buttons, employed gilt toy men and women in ten to twelve different branches. John Baskerville's japanning works in Moor Street exercised a high degree of devolution through managers looking after different departments.

It was not the toy trades, however, but two other Birmingham industries which provided the classic examples of the division of labour analysed in the *Encyclopédie*, Smith's *Wealth of Nations* and Marx's *Capital*. The needle and pin manufactures, though centred in other areas of the country, were carried out also in

Birmingham from the late eighteenth century. The manufacture of needles was divided into eight main operations: pointing, stamping, eyeing, filing, hardening, strengthening, scouring and polishing. A pinworks in Birmingham in 1810 divided the production process into thirteen different operations: the wire was drawn out to a proper size by a simple engine; it was straightened by another machine; it was cut into lengths; the ends of these lengths were pointed by means of a wheel; these were cut again, each wire forming several pins; the wire was twisted by means of a wheel; the twisted wire was cut into heads; the heads were softened in a fire; the heads were put on by hand; the pins were washed; they were then boiled in a liquor of tartar and tin, dried and finally they were papered by children.[17]

Another early Birmingham industry, the nail manufacture, displayed the characteristics of the division of labour in the proto-industrial model. The iron was prepared by rolling heavy bars into sheets; the thin sheets were slit into rods which in turn were rolled into rods of the gauge required for the particular nail. The rods were then cut, headed and pointed by the nailer at his domestic forge. It was an industry carried out in the small-scale family production unit often in some combination with agriculture, but even as early as the seventeenth century it was one of the poorest and most despised of all trades. Nailmaking, in spite of its dispersed and very small production units and the simplicity of the basic processes, was also extremely specialized, but specialized according to product and region. By the early nineteenth century there were twenty nailmaking districts in the West Midlands, each making a different kind of nail or spike. By 1770 there were 10,000 employed in the trade in the Midlands; by 1798 35–40,000. The nailers were exploited by the ironmongers who held them in debt and controlled their sales. As early as 1655 there was a call for the nailers to cooperate in a strike against the 'Egyptian Taskmasters', that is, the ironmongers. And the trade was said to be in a state of decay in 1737, 1765 and again in 1776. Young reported in 1776 that the road from Soho was one long village of nailers who complained that their trade was failing because of disputes with America. When hands were idle they took to other branches, and their children went to Birmingham.[18]

The specialization of firms by product and process was probably the most important determinant of the size and structure of a concern, not just in nailing, but in most of the Birmingham trades. Much of this specialization was, however, ephemeral, for tools and skills could be flexibly turned to whatever happened to be demanded. When bucklemaking started up, the social division of labour was a regional one. The forging was done at Darlaston, the chapes made at Bilston, and the filing, clasping, and putting together carried out in Birmingham. By 1770 nearly fifty of the Birmingham specialist lines were each entered with five or more manufacturers in the business. It is argued that even with the smallest article there was a division between firms that did the final assembly and decoration and other separate firms to provide the raw materials and do the 'stamping, piercing, spinning, brazing, plating and annealing'.[19] The extent of specialization is indicated by over twenty types of buttonmaking being mentioned in *Aris's Gazette* between the 1770s and the 1790s.

By the 1770s specialization became centred on the production of particular groups of articles. Manufacturers decided to produce either brass fittings for furniture or alternatively to provide furniture for new engineering and plumbing trades. Cockfounding became the focus of a group of trades separate from those concentrating on cabinet brass foundry.[20] The glass and toy trades, on the other hand, developed together in Birmingham. Here the glass industry, in contrast to the norm in the rest of the country, started out on a small scale, alongside the manufacture of glass buttons, beads and toys. In fact it was started by small master cutters and toymakers who took over the manufacture of their own raw materials.[21]

Technology

Both the Marxian and the proto-industrial models of manufacture either ignore or give short shrift to the extent of technologi-

cal change to be found in pre-factory industry. Their concentration on the ideal case of the division of labour is
regarded as a substitute for any discussion of technological
change, and there has been little attempt to confront the possible connections between the forms taken by the division of
labour and the types of technological change. Milan Myska's
study of the pre-industrial iron manufacture is one of the few
attempts to look at the impact of a series of technological
changes from the mid-fourteenth to the eighteenth centuries on
the organization of work and to look at these, moreover, not in
textiles but in metal processing. New, more sophisticated technology brought greater specialization of function in the industry,
first developing a clearer division of labour between miners,
charcoal suppliers and foundrymen, and then leading on to
greater specialization among ironworkers themselves – hammer
operations were separated from foundry operations, and new
categories of specialist workers appeared in forging. But this was
under the impact of such major innovations as the application of
water power, the machine hammer, the blow and then blast
furnaces.[22]

Prosser regarded Birmingham as a famous centre for inventors until 1850, pointing out that many more patents were issued
there than anywhere else outside London. This he supposed
was due to the many skilled smiths, founders and engineers in
the town. Most of the patents were granted for small improvements in the manufacture of trinkets and buttons, in machine
tools, and in metal compositions and scientific instruments.
Many such improvements were never patented, but rather just
adopted by the small masters.

The secretive manufacturers who locked their doors and who
led James Drake to complain in 1825 that the tourist trade
was endangered by their behaviour were in all probability
men who found it easier to withhold their innovations by
keeping them dark rather than ensuring the enforcement of a
patent with all the publicity for the specification that this
method involved.[23]

'Turning'
(*Supplement* to *The New and Universal Dictionary of Arts and Sciences*, 1754)

Early recognition of Birmingham's inventive reputation was accorded by the proposal of the Society of Arts and Manufactures in 1761 to award £15 to 'an artist who invented a machine or useful toy thought to occasion a large demand'. The Society made another award in 1763 for the invention of a fine oil varnish used to make paper boxes.[24] Hawkes Smith described the type of invention famous in the town as that which affected 'that alone which requires more force than the arm and tools of the workman could yield, still leaving his skill and experience of head, hand and eye in full exercise'. Because of this Birmingham had supposedly suffered less from the introduction of machinery than where it had been a substitute for human labour. The goods produced by the Birmingham and Midland smithies in the seventeenth century were wrought or cast out of solid metal, so that the earliest working equipment was anvil, hammer, file and grindstone. The lathe succeeded these as the variety of manufacture and the trade in light and fancy articles increased in the later seventeenth century. With the addition of copper, brass and other materials, the rolling mill, stamp, press and drawbench were introduced over the course of the eighteenth century.[25]

Though most of these innovations were tools or mechanical aids to artisan skills, it was also quite likely that scientific knowledge in some form played its part. For it was 'not possible to dismiss the accuracy of the Birmingham pattern maker, moulder or polisher in 1750 as simply the product of experience'. The development of the stamp and press in the 1760s, for example, required a 'high degree of accuracy in the finish of the machine tool, calculations as to stress set up in the metal formed and the degree of leverage required to minimize the operator's effort'.[26]

The stamp, press and drawbench, along with the lathe, were perhaps the most widely used and the most significant of innovations in the Birmingham trades. A great variety of lathes were available by the eighteenth century, though most manufacturers made use only of the simpler round and oval turning lathe. This was operated by a treadle, and produced engine-turned parts in round or oval section. The application of stamping and pressing was made possible by the use of new alloys of copper and zinc. The innovation was first applied to the buckle trade, then to buttons, cloak pins, drawer handles and other toys. Before the seventeenth century metal was simply beaten out with a hammer and the piece shaped into forms by shears:

> to give the impression, the workman with one hand held the die in its proper position on the face of the metal piece, or blank, while with a hammer he gave the blow, which to a certain degree transferred the pattern to the intended coin.

The means of obtaining a perfect impression was acquired in the early eighteenth century with the application of the stamp, or a means of regulating the face of a heavy mass bearing the pattern or die desired. Though invented by a London toymaker in 1766, it was quickly adapted by Birmingham manufacturers to shape hollow ware. The press was devised for cutting out circular plates of sheet metal which afterwards served as blanks to receive impressions from the hammer or stamp. It was also adopted for large medallions and other works of art whose value warranted more time, and whose perfection needed the application of great force. For larger articles, such as medals with two faces, the lever attached to the screw was a bar with heavy weights or a large iron wheel with two handles. The bar or wheel was put into position by two workmen who

> cause it to revolve with great rapidity . . . it makes between two and three revolutions. The screw descending till the die is brought down on the blank with great force . . . The rebound is sufficient with a little assistance to return the apparatus to its original position.[27]

With the development of stamp and press, die sinking became a crucial though not always recognized trade. The die was cast in steel and the design engraved on it. For this the die sinker used gravers of three different shapes, and often had forty or fifty sizes of three types to suit the character of the work. When the engraving was finished the die was heated and suddenly cooled to harden it, then the surface was polished. But the die sinker seldom received any credit for the designs, and indeed it was in the interest of the manufacturer who employed him to suppress his name. G. C. Allen has described the ebb and flow of the die sinker's trade and status in two centuries.

> In the eighteenth century, when dies were first used by the sheet metal trades, the craftsman who made them was, for the most part, trained and employed by the great concerns which had adopted the new method of manufacture. As stamping and pressing spread over the whole range of small metal industries, which were still mainly conducted in workshops, the die sinker became an independent class of craftsmen, working for a number of firms, none of which could keep a die-sinker fully occupied. Then with the rise in the scale of production towards the end of the nineteenth century, the tendency was reversed, and die sinkers again came to form part of the staff of larger factories.[28]

The purpose of the drawbench, the other eighteenth-century innovation of great significance to the Birmingham and other light metal trades such as the pin and needle manufacture, was to elongate a rod or piece of metal, and at the same time preserve an equal thickness throughout by drawing it through a perforation in a hard metal plate.[29]

The effect of these innovations in combination with the division of labour were vividly described in Lord Shelbourne's report on the Birmingham hardware manufactures in 1766.

> Its great rise was owing to two things, first the discovery of mixed metal so mollient or ductile as easily to suffer stamping, the consequence of which is they do buttons, buckles, toys and everything in the hardware way by stamping machines which were before obliged to be performed by human labour. Another thing quickly followed, instead of employing the same hand to finish a button or any other thing they subdivided it into as many different hands as possible. . . .

There are besides an infinity of smaller improvements which each workman has and sedulously keeps secret from the rest. Upon the whole they have reduced the price so now that the small matter of gold on a button makes the chief expense of it. . . . However, they have lately discovered a method of washing them with acqua fortis, which gives them the colour of gold, and are come to stamp them so well that it is scarce possible at any distance to distinguish them from a thread button.[30]

The range of tools and machines owned by most button- and bucklemakers was sufficiently wide to show the impact of technological change on the trades. Buttonmakers typically owned several different-sized stamps a number of setting-out and piercing presses, a variety of different lathes, anvils, bellows, bench vices and various tools. In the case of Ward and Browne, the buttonmakers who sold off their tools and machinery in 1768, along with three stamps they had turning, lap, forging, spinning and die lathes, vices, die punches, sinking tools and a small cutting-out press: other tools included swages, punches, scales and weights, shears and a machine for turning links. James Dalloway's buttonworks in 1778 contained, with other machinery, grinding, turning, finishing, fringing and polishing lathes. Thomas Dawes's small buckleworks in 1772 contained buckle stamps, two large button stamps, a two-sided piercing press, lathes, vices, bellows, a long wheel with frame, and stove plates with a set of casting moulds. A survey of firms going out of business and advertising their capital stock in *Aris's Gazette* over the period 1768–89 indicates that not only buttonmakers, but bucklemakers, toymakers and japanners owned a substantial range of small machine tools like lathes, stamps and presses, as well as numerous and varied tools. These may not represent the entire range of toymakers, for it was also sometimes possible for workmen to rent machinery.[31] But, in general, the variety and extent of machinery and tools in use in the toy trades belie any simplistic notion of garret room or family production as the rule. These Birmingham tools and machines were commonly used in medium-sized firms, and were more or less specialized according to the work undertaken. But equally the type of technology in use was not that adapted to continuous process or mass production. Its hallmark was flexibility and application according to the dictates of artisan skill to the production of a wide range of different articles. While Soho, for example, was employing 800 to 1,000 workers in 1770 along with a wide range of sophisticated machinery, its object was not mass production. New patterns were constantly brought into use, new products devised, and many valuable pieces individually constructed. It was described in the eighteenth century thus:

> The building consists of four squares, with shops, work-houses, etc., for a thousand workmen, who in a great variety of branches, excel in their several Departments; not only in the fabrication of Buttons, Buckles, Boxes, Trinkets etc., in Gold, Silver, and a variety of Compositions, but in many other Arts, long predominant in France. . . . And it is by the Natives hereof, or of the parts adjacent, that it is brought to its present flourishing state. The number of ingenious mechanical contrivances they avail themselves of, by means of Water Mills, much facilitates their work and saves a great portion of time and labour.[32]

This was hand-operated machinery, supplemented in some cases by limited horse or water power. Steam power, though perhaps the most famous product of the town, was hardly ever used there before 1800. By 1815 there were still only forty engines in the town, but a minor proliferation of small steam engines occurred in the 1830s.[33]

If technological change favoured the rise of larger or at least medium-sized firms, so too did the increasing sophistication of mercantile networks. In the early trades these also fitted the pattern of the classic putting-out system. Early on, most master workmen disposed of their goods through a factor who made use of patterns and pattern books with riders sent out to solicit orders and take payments. But Hutton described how this soon changed.

The practice of the Birmingham manufacturer for, perhaps, a hundred generations, was to keep within the warmth of his own forge. The foreign customer applied to him for the execution of orders, and regularly made his appearance twice a year, and though this mode of business is not totally extinguished, yet a very different one is adopted. The merchant stands at the head of the manufacture, purchases his produce, and travels the whole island to promote the sale.[34]

Taylor and Garbett also reported in the mid-eighteenth century that even the large firms were part of a sophisticated indirect mercantile network. Their goods passed through five or six hands – 'Master workmen sell to the Factor, the Factor to the Merchant, the Merchant to the country dealers or shopkeepers in large towns, who sell to the shopkeepers of inferior rank in small towns and villages.' But Boulton made his own business contacts as well as selling through agents. The Boulton and Fothergill agents developed the 'familiar figure of the Birmingham traveller with his set of patterns weighing 5 cwt. by 1836, and forming a full and ample load for a one horse carriage'.[35]

The flexibility of the technology, and the different market and production pressures facing the small, medium and large firms, make it apparent that even in the eighteenth century there were enormous divisions within this so-called ideal of the workshop economy. To argue that the Birmingham economy had an increasing tendency by the nineteenth century to divide into small garret-room family firms does not tell us anything positive about the socioeconomic condition of this economy.[36] For the proliferation of such small-scale units was frequently an expression of depression, or 'industrial involution'. Far from Adam Smith's independent artisan such units were more likely to hold the inmates of Disraeli's Wodgate.

Here Labour reigns supreme . . . the business of Wodgate is carried on by master workmen in their own houses, each of whom possesses an unlimited number of what they call apprentices . . . whom they treat as the Mamelukes treated the Egyptians.[37]

A recent social historian has argued that the model of an artisan economy dominated by skilled labour and absence of large capital investment, and projecting the possibility of upward social mobility, was really inapplicable to nineteenth-century Birmingham. Such a model assumes that the workshop 'escaped the divisive touch of industrial capitalism'.

In Birmingham 'social relationships changed considerably, particularly in the 1830s and 1840s, as competitive production for a mass market altered the face of the workshop economy.' In the first half of the nineteenth century larger establishments were introduced, apprenticeship customs sidestepped and women and children substituted for men. T. C. Salt, a lamp manufacturer, in 1833 defended his new policy of employing women and children:

> formerly when trade was good, we did not resort to that screwing system; if we had done so we should not have had a single workman to work for us the next day.[38]

But even this view assumes that the big divide came in the nineteenth century – that there was still an ideal workshop economy to look back to in the eighteenth. As the evidence presented in this chapter on the scale of firms and the development of technology shows, however, division and capitalist competitive pressure for structural change existed from the very inception of the Birmingham trades. Even in the proto-industrial phase of Birmingham's development, changes in tools, division of labour and markets affected the structure of the enterprise itself. In eighteenth-century Birmingham new technologies and marketing arrangements favoured the rise of the medium-sized firm. The tendency in the nineteenth was for industrialization to take the form of even greater dualism with the rise of both large-scale firms and the proliferation of sweated garret workers. The artisan structure of this town, as of Sheffield and London, was thus neither static nor egalitarian, not in the nineteenth century and not even in the eighteenth.

Further evidence of the divisions between large and small firms is apparent in their different experiences of and reactions to economic recession and changes in the market, as in the 1790s. Both button- and bucklemaking, a mainstay of large and small firms alike from earlier in the eighteenth century, were in crisis as fashions and markets changed. The one faced a fashion change from metal- to cloth-covered buttons; the other a shift in fashion from buckles to shoelaces. Smaller manufacturers tried to fight these developments through petitions. Some larger manufacturers, such as Boulton and Watt, drastically cut back production from the trades and shifted it to engineering, at the same time abandoning large numbers of dependent outworkers and direct employees.[39] Small manufacturers who survived the period only did so by following suit and shifting to new products. One such firm, Kenricks and Bolton, founded to manufacture buckles in 1787, shifted its production within four years. Kenrick set up a new firm in West Bromwich in 1791 to make cast-iron articles; by 1815, drawing on fewer than a hundred indoor and outworkers he was manufacturing a wide range of hardware and hollow ware.[40]

Apprenticeship and labour

Though the tools and machinery in the Birmingham trades 'saved time and labour', it is usually argued that they did not, in the eighteenth century at least, reduce the skill involved in most operations. How, then, was labour recruited to these burgeoning trades and how was it trained? Just as the organization of the trades showed great variety and made use of a technology which fitted the needs equally of small-, medium-, and large-scale firm alike, so the sources and recruitment of labour were mixed. The apprentice, indentured servant and ordinary working hand all existed side by side.

It was often thought that Birmingham 'industry' owed its existence to the lack of guild and apprenticeship restrictions. The city was regarded as 'remarkable in Europe for the in-genuity and multiplicity of its artificers in small branches of the iron manufacture – hardware goods, snuff boxes, buckles and buttons'–.

'Tis further observed that this town, tho' so large and populous, is not a corporation or Borough, nor any monopoly of the particular Branches of Trade, yet so prevalent is the Spirit of Industry, so early learnt their children, and practiced in riper years, that you won't see an idle person lurking about the streets.[41]

But to say that the town was unincorporated did not imply that apprenticeship did not exist. It did, however, mean that apprenticeship took on a very different form from that found in more conventional boroughs such as Coventry. It was argued by William Mayo, for instance, that watchmaking was transferred from Birmingham to Coventry because Coventry's traditional apprenticeship regulations were more amenable to the 'factory system'. This 'factory system'

was first attempted in Birmingham 40 years since [1777] but failed in the undertaking as the method of taking apprentices in that town was different; – one of the manufacturers then removed to Coventry where his partnership has from that time had a large establishment of apprentices – 38 or 40.[42]

Apprenticeship in Birmingham seems to have been a good deal more flexible than elsewhere – in the length of terms, types of training, and opportunities available. Formal apprenticeships might be only one year or much longer, and might require some premium according to the trade and employer. But the demands for those with any training and the number of reports of runaway apprentices indicate that the completion of a formal apprenticeship was no requirement for entry to the trades. There were complaints that masters sometimes took on as many apprentices as possible as a form of cheap labour and to avoid employing journeymen. This was said to be especially the case in the japanning and painting businesses in the 1760s.[43] But conversely many employers did not bother with apprenticing their labour. Of 108 buttonmakers in Birmingham in 1767 only 18 took indentured apprentices or had themselves served as apprentices. John Taylor, for example, employed 500 workers, but none of these was serving a formal apprenticeship. There was widespread use of women's and children's labour. In Taylor's button factory women were supposed to have preponderated, and the coining staff at Soho employed 13 men, 27 women, and 16 boys.[44] The flexibility of apprenticeship arrangements in the new trades seems also to have affected the older trades. Threats of prosecution were made in 1777 against stirrupmakers in the town who carried on the trade without having served an apprenticeship.[45]

A mixed labour force of apprentices, hands, and women and children received a training in the basic and varied skills which could be used in many branches of metalwork or finishing and painting. John Fox, who was apprenticed to a toymaker in 1741, was listed as a jeweller in 1767. A runaway apprentice, James Knott, in 1750 was described as capable of working at painting, engraving, plating of spurs, and writing a good hand.[46] The varied application of this training in spite of the rapid differentiation of many of the trades was something which certainly benefited the jewellery trades. These expanded on the basis of labour which could be diverted from the buckle and toy trades once the demand had induced engravers, die sinkers, solders, platers, polishers and others to turn to the manufacture of jewellery. There were some trades, however, such as plated ware, in which specialization did matter. Where Sheffield concentrated on the heavier plated work and Birmingham on the smaller and more high-class work, there was little transferability of labour between the two branches. Though labour could be transferred between saddlery plating and other plating, this could not be done in other branches. James Ryland, a plater of coach harness and saddlery furniture, claimed before the 1812 Committee that in saddlery all the plating was done in Birmingham. 'This description Sheffield can't do. . . . Their plating is entirely different . . . a great deal of the metal plated at Birmingham is used in Sheffield.'[47]

It is argued that there were few periods of general unemployment or underemployment in eighteenth-century Birmingham, for by 1750 the diversity of the trades was such that a temporary setback in one was compensated within another. The distress that did follow the demise of the buckle trade was soon relieved by the rapid rise of the button manufacture and the easy transfer of skills from the one to the other. The series of advertisements for apprentices and journeymen as well as reports of runaway apprentices in the toy, button and japanning trades between the 1760s and 1790s was a positive indication of the high demand for labour and the easy availability of alternative employment. Wages were also high. Arthur Young reported that labour in the surrounding countryside was paid at about £15 a year. At the same time no adult labourer in Birmingham was earning less than 7s a week and some were gaining £3. The average rate for women workers was 7s per week, and for children 1s 6d to 4s 6d. Male workers in the button trade could often gain 25s or 30s per week.[48] These wage rates cannot, however, be taken entirely at face value, for apart from the wage there were customary rights and fringe benefits on the one hand, and outpayments to apprentices, rent for bench space and 'discounts' on piece prices on the other.

The work was also executed, in the main, without the direct supervision of the merchants. Though workers were frequently controlled indirectly by merchants through debt and access to the market, they were not supervised by them. In the nail industry the smiths occasionally had bargaining power, and where they could they played off one master against the other. And bad workmanship, such as sending off 'nails' without heads, was a ready answer to inadequate remuneration.[49]

The opportunities for embezzlement and fraud inherent in the 'manufacturing' merchant's lack of control over the work process were also frequently taken up in the toy trades. Reports of theft of silver and brass were made many times in *Aris's Gazette* between the 1740s and 1790s. The theft of materials was so widespread even in the 1750s that a bill was proposed for 'preventing stealing and buying and receiving stolen lead, iron, copper, brass, base metal and solder'. Buckle blanks cast for stamping buckles were frequently reported stolen. And there were numerous complaints of mixing metals with gold and silver in toys and snuff boxes, falsely claiming the production of

Emailleur à la lampe, perles fausses
(*L'Encyclopédie Diderot*)

gilt and plated ware, and manufacturing illegally covered buttons.[50]

Douglas Hay has recently discussed the extent, the form, and the prosecution of embezzlement and industrial theft in the Black Country. He argues that there were three different kinds of metal theft: embezzlement in workshops, pilfering or robbery of metals for sale or for raw materials for an artisan-thief, and theft of finished products. The complexity of relations among artisans, middlemen and masters frequently made the line between embezzlement and theft difficult to determine. The line between embezzlement and custom was also often unclear, thanks to the widespread practice of allowing workmen with their own tools to take filings from the work as part-payment. The many small shops in the enormous variety of trades in the area provided a ready market not only for such filings but also for a large amount of stolen metal impossible to trace once melted down. Selling and receiving were much safer in the case of unworked metal than of finished goods which could often be traced to an area or even to a particular artisan by the peculiarities of workmanship. Midway between this was the theft of unfinished work, to be finished off in one's own shop and then sold as one's own work. Files, chapes and nails were frequently stolen and disposed of in this manner.[51] Embezzlement was endemic to the small-shop type of work organization found in the hardware trades; it formed one of the major discipline problems endemic to the system of manufacture and to proto-industry.

Another characteristic inherent to the manufacturing mode of production was the hierarchical division of labour. Did the new technologies and new processes developed in the Birmingham trades over the eighteenth century entail the use of more unskilled labour?

Evidence that the Birmingham toy trades made use of apprenticed, nonapprenticed and women and children's labour from the start in many of the processes makes it difficult to assess the impact of new technology on skills and the status and structure of the labour force. Apprenticeship and the sexual division of labour were not necessarily any index of changes in the labour process. Certainly the new technology did affect the division of labour in the trades. Contemporaries argued simultaneously that the techniques used in Birmingham did *not* reduce the skill or labour required in production processes, and that the new machinery allowed extensive use of child labour. Taylor and Garbett, for instance, reported that Birmingham machines reduced the manual labour and enabled boys to do men's work. Shelbourne cited a division of labour that made work so simple that 'five times in six, children of 6 or 8 years old do it as well as men'. And Dean Tucker described the close connection between machinery and child labour in Birmingham. 'When a Man stamps on a metal Button by means of an Engine, a Child stands by him to place the Button in readiness to receive the Stamp, and to remove it when received and then to place another.'[52] But equally, the differentiation of existing trades and the proliferation of a whole number of new trades reflected changes in product as well as in processes. It is particularly difficult to gain any clear idea of the sexual division of labour in trades which displayed such varied industrial structures. It is said, however, that the adoption of machines for stamping and piercing extended the range of female employment especially for young girls.[53] And it was recognized that women's work was widespread in the japanning and the stamping and piercing trades. Girls were specifically requested in advertisements for button piercers, annealers, and stoving and polishing work in the japanning trades. Another advertisement for button burnishers in 1788 also sought 'a woman that has been used to looking over and carding plain, plated and gilted buttons, also a few women that have been used to grind steels, either at foot lathes or mill'.[54] In the nineteenth century the tools were fitted into the press by male toolworkers who also attended to the condition of the tools in cutting the larger stamped work. But women worked even with the large presses, though girls were left to cut out smaller examples of the work. When women were employed in piercing and cutting- out work, they received only 8s–12s a week and girls got 6s–8s; while the toolmaker who superintended the work claimed 30s–40s.

The delicacy of the work in buttonmaking and piercing, as well as in the handpainting of designs, was regarded as the special province of women and girls with their smaller hands and the deftness and concentration already acquired at household needlework. The lacquering and japanning trades required stove management and even in the nineteenth century it was women who worked in the trade. The small lacquering rooms in the brass trades, only 12 by 15 feet and 11 feet high, characteristically contained a couple of iron plate stoves and five to six women workers.

In the nineteenth century women were still employed over a wide range of processes in the Birmingham trades, but these were by and large concentrated in the newer, lighter or more unskilled branches. Women did the lacquering in the brass shops, japanning in the tin plate ware manufacture, and barrel boring in the gun trade. In the button trade there was a division between the old and more skilled branches such as the metal and pearl button section which employed men, and the new covered and linen button section, which employed women. Men made high-class jewellery, and women and girls were left to the cheap end of the trade in gilt articles and chains.

In the Black Country trades there was no lighter alternative work open to women, and they worked beside the men in heavy industry – on the pitbank, in the nail manufacture, and in the manufacture of chains, saddlery, harness and hollow ware.[55] But in many of these trades, and in particular in nailmaking, they had long been degraded workers.

The most celebrated women workers of the West Midlands trades were the nailers. Their subservience in this degraded and poverty-stricken trade reflected the wider subservience of their sex. William Hutton, on his travels in 1741, provided an exemplary male image of this workforce.

> In some of these shops I observed one, or more females, stript of their upper garment, and not overcharged with their lower, wielding the hammer with all the grace of the sex. The beauties of their face were rather eclipsed by the smut of the anvil; or in poetical phrase, the tincture of the forge had taken possession of those lips, which might have been taken by the kiss. Struck with the novelty, I enquired, 'whether the ladies of this country shod horses?' but was answered, with a smile, 'they are nailers.'
>
> A fire without heat, a nailer of a fair complexion, or one who despises the tankard, are equally rare among them.[56]

Yet women appeared to have a knowledge of a wide range of trades, if widowed tradeswomen can be used as evidence. Many women carried on with their husbands' businesses after their death, and though they may have employed some journeymen they would themselves have had to have a great deal of practical experience and knowledge to make a success of running what were in the main small artisan businesses. These women ran the businesses where one might have expected women's work – as in toy-, button- and bucklemaking, and japanning. But widows and daughters also appeared in strength in the iron business, in plumbing and glazing, in the brass-founding and pewtering trades, and among the hammer-, anvil- and edgetool making trades. A survey of *Aris's Gazette* from 1752 to 1790 indicates that women were taking over husbands' businesses or dealing with various problems which arose in the trades over a wide range of processes. Notices appeared from nine female ironmongers, eight plumbers and glaziers, seven buttonmakers and seven bucklemakers, six watchmakers, five brass manufacturers and braziers, five toolmakers, and five chain- and toymakers. There were notices from three women running ironworks, three female plateworkers, two nailworkers, two women running coalworks, as well as individual locksmiths, japanners, wireworkers, and file cutters.[57]

Women occupied an important place in the Birmingham toy trades, as workers and employers. Though the evidence available does not indicate the extent to which there was a 'sexual division of labour' between individual trades and processes, it does indicate an economic and social subservience to men, for their wages were much lower, and they appear as tradeswomen and owners of businesses in their own right in effect only where they were continuing the business of a deceased husband or father. But we cannot deny the knowledge and expertise possessed by such women in these trades, for their businesses were mainly small-scale or at most medium-scale enterprises. And success for women as much as for men in these Birmingham trades was dependent on skill and on knowledge.

Conclusions

To conclude: technology, industrial organization, and the structure of the workforce did not in the case of the Birmingham trades fit any of the ideal types of pre-industrial manufacture analysed in the Marxian or proto-industrial models. Their experience conveys no sense of the rigidities or one-way connections between particular types of technology and the specific forms of industrial organization which we traditionally associate with the period before the factory – before the Industrial Revolution.

However, the experience of the Birmingham trades should not be regarded as a special case. Many of them never did make the transition to the factory system; others degenerated to sweated trades. When steam power was adopted in many branches of the trades in the nineteenth century, it did not transform the labour processes, but was simply incorporated into existing technologies. The use of steam power in Birmingham did not necessarily entail large factories and capitalist control or supervision. Such power was frequently rented out in large buildings full of small workrooms sublet to individual artisans. A working man could continue to use the same basic tools developed in the eighteenth century and rent room and power to pursue his trade at greater speed and efficiency. But steam power imposed a much greater regularity on the working day, and even the self-employed artisan could no longer organize his working day around his other familial, cultural and community commitments.[58]

Many other eighteenth- and nineteenth-century industries shared this experience. They did not fit the ideal types of the models, but they did experience a transition to modern industry in the nineteenth century. The manifestation of this transition was, however, as complex and variable as 'manufacture' and 'proto-industry' had been in the seventeenth and eighteenth centuries.

Conclusion

The age of manufactures in Britain was a complex web of improvement and decline, large- and small-scale production, machine and hand processes. This book has attempted to present some of the richness and variety of early industrial Britain. It has explicitly redressed the balance of recent teleological accounts of the process of industrialization. It has abandoned the perspective which probes back to the eighteenth century for examples of the 'modern', for instances of striking increases in productivity, and for some clearly defined path into nineteenth-century industrial greatness. It has challenged the applicability of the economists' growth models and stage theories, which have narrowed our account of historical processes to aggregate and macroeconomic analysis. Such purely economic history has contributed little to understanding the wider historical framework, especially social history. In many ways, purely economic history has presented a misleading picture of the eighteenth-century economy itself: focusing on but one path into industrialization, it has cut us off from understanding alternative paths, and the whole experience of traditional and declining sectors.

My book has drawn on a variety of sources and forms of argument from economic, social and cultural history in order to present a more historical and less instrumental picture of eighteenth- and early nineteenth-century industry. The picture is incomplete: I have been able to give little space to several industries – building, mining, leatherworking, food processing – which deserve more prominence. Here, as in more conventionally economic histories, the focus has been on the two industries which dominated the foreground: textiles and metals. It is precisely through detailed focus on these two best-

known cases, however, that I have sought to deepen the conventional perspective, amplifying the presentation of the
economic, social and cultural experiences which lay behind the
industries.

The book's argument has proceeded at a macro and a micro
level. The first part of the book discussed the structure of
industrial growth and its relation to the wider economy and
society. Successive chapters examined perceptions of industry in
the eighteenth century, the relations of industry to the
agricultural sector and the experiences of regional decline. This
part included an assessment of the two most fertile of the
available theoretical frameworks for understanding eighteenth-
century industrial growth – the theory of proto-industrialization
and Marx's theories of primitive accumulation and the phase of
manufactures. This assessment suggested an urgent need to go
beyond the rigid frameworks of both to include culture and
community in our explanations of industrial and technical
change. At the micro level, the second part of the book
described and compared the variety of forms of industrial
structures and technologies, and their very different outcomes.
Through a combination of extended narrative and socioeconomic analysis the experiences of the textile and metal
industries were examined afresh, and set off against the
economic theory of technological change and the Marxist
analysis of the labour process.

My picture of the age of manufactures has, I believe, high-
lighted the following points. First, industrial growth took place
over the whole of the eighteenth century, not just in the last
quarter of it. There was substantial growth in a whole range of
traditional industries as well as in the most obviously exciting
cases of cotton and iron. Second, technical change started early
and spread extensively through industry. Innovation was not
necessarily mechanization. It was also the development of hand
and intermediate techniques, and the wider use of and division
of cheap labour. It was above all a conjuncture of old and new
processes, and that conjuncture affected performance and work
experience. Thus, third, industrialization was about work
organization; decentralization, extended workshops, and

sweating were equally new departures in the organization of production. There was no necessary progression from one to another; their relative efficiency depended on the economic context, and almost any combination of them was possible. These industrial forms, furthermore, had their origins in the differences in work organization to be found in industries like metals and textiles from the very early industrial or proto-industrial phases. Marked differences in production arrangements existed from the beginning of the eighteenth century among the wool, worsted, knitting, silk, linen and cotton industries. In turn, these differences were rooted in socioeconomic structures: historically established levels of industrial concentration and social inequality, legal and customary regulations, and regional cultural traditions.

Fourth and finally, but not least, my book has demonstrated the variable impact of technical and industrial change on the division of labour, skills, employment and regions. It has shown that such change did not always and everywhere entail growth. On the contrary, regional industrial decline affected large numbers of workers, and the traditional industries practised in areas of decline, especially the South, were among the most important employers of labour in the eighteenth century. The eighteenth century was no 'golden age' for the labourer, and though decentralized processes in cottage or workshop manufacture predominated, much of this manufacture was marked by poverty and insecurity. Growth itself, moreover, need not benefit labour. The expansion of the eighteenth-century textile and metal industries depended on the recruitment of huge quantities of cheap female and child labour. This book has shown how some technologies and processes were adapted to the use of large quantities of cheap labour. But this was often the same labour which subsequently suffered the first widespread dislocation from technological change, and mounted the first great waves of resistance to machinery.

These four conclusions seem to me to point towards two directions for future research. I would suggest first that in moving away from our current narrow vision of industrialization, we take research beyond the conventional territory of cotton, iron and steam power. Fewer people, less wealth were tied up in those technological leaders than in the traditional textile industries and metal manufactures, and this book has sought to give due attention to the range of woollen, linen, silk, cutlery and hardware trades. But large numbers were also to be found in mining, building and the food, drink and leather trades; and there were also the luxury and service trades of town and village. Existing research on these is limited, and analysis in the broad socioeconomic terms here outlined has been virtually nonexistent. Accordingly, they have yet to be given their historical due. This book itself has continued the conventional focus on the textile and metal industries, but it has sought to give more weight to their range. We need now to move further out, to discover the processes, organization and labour forces of many more eighteenth-century industries.

Second, there is an urgent need for enquiry into the social realities behind 'increases in productivity'. It is beyond doubt that the technical, organizational and structural changes behind increases in labour productivity involved the dislocation of labour and communities. Hand in hand with industrialization went acute divisions in social experience. One of the most pervasive divisions was that between regions; another was the division of labour, especially by gender. Our curiosity must surely be whetted by the very different responses of workers to technological change in the eighteenth century. Some were enthusiastic or at least passive; others entirely negative. And the reasons for this seem to lie to a large extent in the regional pattern of industrial growth and decline. For the reception accorded technological change in regions where industry had ceased to grow, or where social inequality was marked, was much more negative than in regions of economic opportunity and more egalitarian social structures. The new textile machinery was fiercely resisted in the South, and especially in the West Country, while there seem to be no major instances of resistance to the introduction of machinery in Birmingham. There was more to it than simply differences in regional economic opportunity and social structures, however. Resistance to textile machinery was found in the developing North, even in the cotton districts; and almost certainly this has to be explained in terms of community and cultural traditions along with the emergence of new social divisions. This book has drawn attention to the potential role of artisan organization and especially the solidarity of the outworker communities in the reception to technical change: both of these need to be explored in further comparative research.

Still more in need of investigation is the division between male and female labour markets. In the nineteenth century the great public debates on machinery allowed the voices of artisans and sweated handloom weavers, many of them urban male workers, to be heard. But the much earlier, eighteenth-century voices of protest against the machine, voices from the country-side and especially from women, were drowned out at the time by the proclamations of the 'improvers', and have since been largely ignored by historians. My book has revealed something of the reception which women domestic workers gave to machinery, and of the complexity of their responses; both favourable and hostile. But we need to know much more about the extent to which there were 'women's technologies', and the extent to which there was a 'gender bias' in technological development. We need to know too about the behaviour of women workers in family and community work settings. What part did their work patterns and social networks play in determining the structure of work organization and the reception to new technologies?

Above all else this book indicates the need for a reconciliation of social and economic history. For too long their divorce has entailed a major gap in our understanding of the deep social divisions which accompanied industrialization, and which are still so evidently present at the heart of the current decline and restructuring of the British economy.

Notes

Introduction

1. See, for example, the differences between the definitions put by Arnold Toynbee, who coined the term – 'The essence of the Industrial Revolution is the substitution of competition for the medieval regulations which had previously controlled the production and distribution of wealth' (Toynbee, *Lectures*, 85) – and the definition in Peter Mathias's textbook – 'The concept implies the onset of a fundamental change in the structure of an economy; a fundamental deployment of resources away from agriculture' (Mathias, *The First Industrial Nation*, 2). For further discussion of economic historians' problems in defining both the Industrial Revolution and technology, see Michael Fores, 'The Myth of a British Industrial Revolution', and 'Technical Change and the Technology Myth'.
2. See, for example, Floud and McCloskey, *Economic History of Britain*.
3. Particularly fine examples of such work on other countries are Reddy, 'Skeins, Scales, Discounts'; Sider, 'Christmas Mumming'. Other historians have explored these themes in later nineteenth-century Britain. See Price, *Masters, Unions and Men*; Lazonick, 'The Case of the Self Acting Mule', 231–62. There are only a few general surveys touching on these questions for the eighteenth century. See Rule, *Experience of Labour*, and Malcolmson, *Life and Labour*.
4. Froebel, Heinrichs and Kreye, *New International Division of Labour*; Pearson, 'Reflections on Proto-industrialization'.
5. Sabel and Zeitlin, 'Historical Alternatives to Mass Production'.
6. Goody, *From Craft to Industry*; Schmitz, *Manufacturing in the Backyard*; Pearson, 'Reflections on Proto-industrialization'.
7. Ashton, *Economic History of England*; Mantoux, *Industrial Revolution in the Eighteenth Century*; C. Wilson, *England's Apprenticeship*.
8. Usher, *Industrial History of England*; Unwin, *Guilds and Companies*

of London; W. Cunningham, *Growth of English Industry*; Marshall, *Industry and Trade*, Appendix B. See Kadish, *Oxford Economists*, and Maloney, 'Marshall, Cunningham and the Emerging Economics Profession', for a discussion of the school of historical economics in England. See Kriedte, Medick and Schlumbohm, *Industrialization*, for a discussion of the historical school and its aftermath in the wider European context.

9. Clapham, *Economic History*; the Hammonds, *Rise of Modern Industry*.
10. Clark, *Working Life of Women*; Pinchbeck, *Women Workers*; George, *England in Transition*, and *London Life in the Eighteenth Century*.
11. See Schmiechen, *Sweated Industries*; and Bythell, *Sweated Trades*.

1. Industries

1. Landes, *Unbound Prometheus*, 41.
2. ibid, 42.
3. ibid, 43.
4. O'Brien and Keyder, *Economic Growth in Britain and France*, 57, 61, 62; Crafts, 'British Economic Growth', 187.
5. Kuznets, *Modern Economic Growth*, 64.
6. McCloskey, 'The Industrial Revolution', 118.
7. J. G. Williamson, 'Why Was British Growth So Slow during the Industrial Revolution?' Harley, 'British Industrialization before 1841'; Crafts, 'British Economic Growth'.
8. Tables 2, 3 and 4 are from Deane and Cole, *British Economic Growth*, 78; Crafts, 181,185.
9. Lindert, 'English Occupations 1670–1811', 702–5, and 'Revising England's Social Tables 1688–1812'.
10. Deane and Cole, 52.
11. Landes, 42.
12. Deane and Cole, 52.
13. Darby, *A New Historical Geography*, 56–7.
14. Deane and Cole, 185, 187, 196; Wilson and Parker, *Sources of European Economic History*, 124.
15. Deane and Cole, 185, 187.
16. Harte, 'The Rise of Protection', 104; Durie, 'The Linen Industry', 89; Deane and Cole, 51.
17. Mitchell and Deane, *Abstract of British Historical Statistics*.

18. Harte, 104.
19. Deane and Cole, 51.
20. idem.
21. ibid, 55.
22. Darby, 67.
23. Deane and Cole, 51.
24. Mathias, *First Industrial Nation*, 482.
25. Darby, 67.
26. Landes, 95.
27. Mathias, 466.
28. Cited in Darby, 69.
29. Crafts, 180.
30. See, for example, Mathias, *The Brewing Industry in England 1700–1830*, and Coleman, *The British Paper Industry 1495–1860*.
31. Gatrell, 'Labour, Power and Size', 125.
32. Chapman, 'Industrial Capital', 113–38.
33. Kuznets, chap. 3.
34. idem.
35. Deane and Cole, 142. (Crafts has revised these sectoral weights, giving somewhat greater weight to industry and commerce, and smaller to agriculture: Crafts, 189.)
36. idem.
37. Von Tunzelman, 'Technical Progress', 161.
38. Lindert and Williamson, 'English Workers' Living Standards', 12.

2. Political Economy and the Growth of Manufacture

This chapter is closely based on my article 'Political Economy and the Principles of Manufacture 1700–1800', in M. Berg, P. Hudson and M. Sonenscher, *Manufacture in Town and Country before the Factory* (Cambridge, 1983). I would like to thank Cambridge University Press for permission to print a revised version here.

1. Cited in George, *England in Transition*, 29.
2. See Berg, *Machinery Question*, 1–8.
3. Appleby, *Economic Thought and Ideology*, 156–70.
4. McCulloch, *Early English Tracts*, xii, xiv. The tract was not, however, in Adam Smith's library.
5. Martyn, 'Considerations', 569.

6. ibid, 586, 589, 590.
7. ibid, 590–1.
8. ibid, 615.
9. For a discussion of the effects of the changing economy on economic ideas in the eighteenth century see Coates, 'Changing Attitudes'.
10. Cary, *Essay on the State of England,* and *Essay Towards Regulating the Trade.* Appleby lumps Cary with the group of free traders even though his arguments favoured protection for home industries and he condemned the East India trade.
11. Gee, *Trade and Navigation,* 5, 69, 70.
12. Defoe, *Tour,* 493–4.
13. ibid, 496.
14. George, 46–7.
15. Tucker, *Instructions for Travellers,* 23–4.
16. Thirsk, 'Fantastical Folly of Fashion', 71.
17. Mann, *Cloth Industry in the West of England,* 102, 104.
18. *Reflections on Various Subjects.*
19. Tucker, *A Brief Essay,* 27.
20. *Reflections on Various Subjects,* 24.
21. ibid, 24.
22. Postlethwayt, *Britain's Commerical Interests,* II, 416, 420.
23. ibid, 420.
24. A. Anderson, *Historical and Chronological Deduction of the Origin of Commerce*; Kenrick, *Artists and Manufacturers of Great Britain.*
25. J. Cunningham, *Essay on Trade and Commerce,* 92.
26. ibid, 130.
27. Smith, *Lectures on Justice,* cited in *Wealth of Nations,* Vol. I, Book I, chap. ii, 31ff. All subsequent references are from Vol. I.
28. Smith, *Wealth of Nations,* I, ii, 31.
29. ibid, I, viii, 104.
30. ibid, III, iii, 409.
31. ibid, 409.
32. ibid, 101.
33. ibid, I, viii, 101.
34. ibid, 83.
35. ibid, III, iv, 422.
36. Thirsk, *Economic Policy and Projects,* 150–1.
37. Smith, *Wealth of Nations,* IV, viii, 644.
38. J. Anderson, *Agriculture, Commerce, Manufactures and Fisheries of Scotland,* I, 39, 53.

39. Cited in Stewart, *Lectures on Political Economy, Collected Works*, VIII, 164
40. ibid, 165.
41. ibid, 175–6.
42. ibid, 177.
43. Cited in O'Brien, *J. R. McCulloch*, 283, and McCulloch, 'On Cottage and Agrarian Systems', 41.
44. Mill, *Principles*, 125 and 127.
45. Engels, *Condition of the Working Class*, 37.
46. See S. C. on Woollen Industry, *P.P.* 1806, cited in Clayre, *Nature and Industrialization*, 66.

3. Models of Manufacture

1. Marx, *Capital*, I, 875.
2. Thorold Rogers, *Six Centuries of Work and Wages*, 489–90.
3. Marx, *Capital*, I, 912.
4. ibid, 928.
5. ibid, 911.
6. ibid, 1026.
7. ibid, 465.
8. Marx, *Grundrisse*, 510–11.
9. Myska, 'Pre-industrial Ironmaking', 45–7.
10. 'The Babbage Principle' in Babbage's own words was 'that the master manufacturer by dividing the work to be executed into different processes, each requiring different degrees of skill or of force, can purchase exactly that precise quantity of both which is necessary for each process, whereas if the whole work were executed by one workman, that person must possess sufficient skill to perform the most difficult and sufficient strength to execute the most laborious, of the operations into which the work is divided'. See Berg, *Machinery Question*, 182–9.
11. Marx, *Capital*, 489–90.
12. ibid, 490.
13. See Diderot, *Encyclopédie*, vol. 5, 'Epingle'.
14. Lord Fitzmaurice, *Life of Shelbourne*, 404.
15. Roll, *An Early Experiment*.
16. Chapman and Chassagne, *European Textile Printers*, 97.
17. Marglin, 'What Do Bosses Do?', 20; The Brighton Group, 'Capitalist Labour Process'; Berg, *Technology and Toil*.

18. Dobb, *Studies*, 143 and chapter 4.
19. Mendels, 'Proto-industrialization'; Medick, 'The Proto-industrial Family Economy'; Jones, 'Agricultural Origins'; de Vries, *Economy of Europe*, 95–6. Some of the discussion of proto-industrialization in this section of the chapter is based on Berg, Hudson, Sonenscher, 'Introduction', *Manufacture in Town and Country*, and I am indebted to my co-authors.
20. Habakkuk, 'Population Growth'; John, 'Agricultural Productivity'; Thirsk, 'Industries in the Countryside'; Chambers, 'Vale of Trent'; Chambers, 'The Rural Domestic Industries'; Jones, 'Agricultural Origins'.
21. Mendels and Deyon, 'Theory and Reality'; Coleman, 'A Concept too Many'.
22. Jeannin, 'Développement ou Impasse?', 52–65; Kriedte, Medick, Schlumbohm, *Industrialization*, 7; Houston and Snell, 'Proto-industrialization'.
23. Schlumbohm, 'Productivity', 4.
24. Medick, 'Proto-industrial Family Economy', 301–10.
25. Jones, 'Environment', 494.
26. Medick, 'Proto-industrial Family Economy', 301.
27. Medick, 'Plebeian Culture', 89–92.
28. O. E. Williamson, 'The Evolution of Hierarchy', and Millward, 'The Emergence of Wage Labour'.
29. Clapham, *Economic History*, I, 145, 191; Hudson, 'From Manor to Mill'.
30. Faucher, cited in Clapham, ibid, 175.
31. Schremmer, 'Proto-industrialization', 123.
32. DuPlessis and Howell, 'Early Modern Urban Economy', 51 and 84.
33. For a discussion of these artisan structures in an unincorporated town such as Birmingham see Chapter 10. For further discussion of these artisan systems in Britain and France see Sabel and Zeitlin, 'Historical Alternatives'. Artisan organization and customs in Britain and France are discussed in Prothero, *Artisans and Politics*; Rule, *Experience of Labour*; Sewell, *Work and Revolution*.
34. Thompson, *Making of the English Working Class*, 13.
35. Clapham, *Economic History*, I, 156.
36. These are described in Chapman, 'Industrial Capital', 124.
37. Chapman and Chassagne, 215, 194. Also see Freudenberger, 'Protofactories', for a discussion of the characteristics of this industrial structure and how it was developed on the landed estates of the Hapsburg Empire in the eighteenth century.

38. See Sabel and Zeitlin for the different success rates by which three artisan regions adapted to changing markets. See Thomson, 'Variations in Industrial Structure'.
39. Coleman, 'A Concept too Many', 445.

4. Agricultural Origins of Industry

1. Mill, 'Nature, Origins, and Progress of Rent', 177.
2. Jones, *Agriculture*, chapter 3; Thirsk, *Agrarian History*, Introduction.
3. Deane and Cole, 15.
4. Wrigley, 'A Simple Model of London's Importance'.
5. Wrigley, 'Growth of Population', 122; Deane and Cole, 45; Mitchell and Deane, 8.
6. John initially made the case for a connection between low agricultural prices and rising industrial demand. Recent work throws this into question. See Cole, 'Factors in Demand', and Beckett, 'Regional Variation'.
7. Gould, 'Agricultural Fluctuations'.
8. Eversley, 'The Home Market', 206–59.
9. Jones, 'Agricultural Origins of Industry', 138.
10. These arguments are made by Hohenberg, 'Toward a Model of the European Economic System'.
11. Jones, *Agriculture*, 117.
12. Hoskins, *The Midland Peasant*, 269.
13. Eden, *The State of the Poor*, I, 555.
14. See Westerfield, *Middlemen*, for a survey of the distinctions between the various functions of the corn middlemen. Also see Thompson, 'The Moral Economy', for a discussion of monopolies in the retailing of corn.
15. I owe this point to Jeanette Neeson. Also see Beckett, 'Regional Variation', 41–2.
16. Cole, 54–7.
17. Wrigley, 'The Growth of Population', 97–8.
18. Jones, *Agriculture*, 102–10 produces the usual examples of landlords investing in mining and ironworking on their estates. But the motivation behind their industrial investment may have been as dubious as that behind their investment in the land itself. For a discussion of this see Cooper, 'In Search of Agrarian Capitalism'.
19. B. L. Anderson, 'The Attorney'. Also see Rowlands, 'Society and

Industry in the West Midlands', who shows from local studies that credit was available to all but the lowest levels of society, and that anyone with any land at all could raise a mortgage. Smiths and husbandmen could raise £20 to £40 in this way.

20. Burley, 'An Essex Clothier'; Ashton, *Peter Stubs.*
21. Chapman, 'Industrial Capital'.
22. The classic statement of this position is Chambers, 'Enclosure and Labour Supply'.
23. Hoskins, *The Midland Peasant*, 263.
24. Jones, 'Agricultural Origins', 139.
25. For a standard statement of this position see Bythell, *Handloom Weavers.*
26. I owe this point to J. Neeson.
27. Thirsk, 'Industries in the Countryside'. This paragraph is based on arguments made in Berg, Hudson, Sonenscher, 'Introduction', *Manufacture in Town and Country*, 22–3.
28. ibid. Also see Hey, 'A Dual Economy'.
29. Chambers, 'The Rural Domestic Industries', 430.
30. Jones, 'Agriculture and Economic Growth', 110–11.
31. Spufford, 'Peasant Inheritance Customs and Land Distribution', 157.
32. Howell, 'Peasant Inheritance Customs in the Midlands 1280–1800'.
33. For an early statement of this see Habakkuk, 'Family Structure'.
34. Hey, *Rural Metalworkers.*
35. Thompson, 'The Grid of Inheritance'.
36. Houston and Snell, 'Proto-industrialization'.
37. Rowlands, *Masters and Men*, 39–43.
38. See Hudson, 'Proto-industrialization'.
39. The extent of local industrial specialization within regions is noticed by Sydney Pollard in his *Peaceful Conquest*, 32–5, but no attempt is made to link these specializations to their individual agrarian contexts.

5. Industrial Decline

1. Thirsk, *Economic Policy*, 109. Also see Thirsk, 'Industries in the Countryside', and 'The Fantastical Folly of Fashion'.
2. Thirsk, *Economic Policy*, 168.
3. Wrightson, *English Society*, 139.
4. Pollard, 'Industrialization and the European Economy', and *Peaceful Conquest.*

5. Richards, 'Women in the British Economy', 343.
6. Jones, 'Constraints on Economic Growth', 423–30, and Jones, 'Environment'.
7. Polanyi, *The Great Transformation*.
8. Little, *Deceleration*, 20–1.
9. Jones, 'Constraints', 423.
10. Defoe, *Tour*, 464.
11. Wrightson, *English Society*, 139.
12. Hoskins, *Midland Peasant*, 274.
13. Jones, 'Constraints', 425.
14. Pollard, *Peaceful Conquest*, 14.
15. Jones, 'Constraints', 429; Pollard, *Peaceful Conquest*, 20.
16. Jones, 'Constraints', 426.
17. Cited in Brown, *Essex*, 1.
18. Brown, ibid, 14. Also see Coleman, 'Growth and Decay: the Case of East Anglia'.
19. Defoe, *Plan of English Commerce*, 257, cited in George, *England in Transition*, 55.
20. Lloyd Prichard, 'The Decline of Norwich', 373, 374; Coleman, 'Growth and Decay'.
21. R. G. Wilson, 'The Supremacy of Yorkshire', 231–7.
22. Lloyd Prichard, 375.
23. Mann, *Cloth Industry in the West of England*, 159–76.
24. R. G. Wilson, 233.
25. Jones, 'Agriculture and Economic Growth', 111.
26. Brown, *Essex*, 19–25.
27. See previous chapter and Chapman, 'Capital Formation'.
28. Jones, 'Agriculture and Economic Growth', 105.
29. Jones, 'Environment', 498.
30. Tucker, *Instructions for Travellers*, cited in R. G. Wilson, 238.
31. R. G. Wilson, 238.
32. Mann, *The Cloth Industry*, 97–9, 192.
33. R. G. Wilson, 241.
34. Brown, *Essex*, 20, 25.
35. Mann, *The Cloth Industry*, 114.
36. ibid, 114, 125, 126.
37. Jones, 'Constraints'; Mann, *The Cloth Industry*, 161, 149.
38. Pinchbeck, *Women Workers*, 155–6.
39. I owe this point to J. Neeson.
40. Brown, *Essex*, 113, and Burley, 'An Essex Clothier', 289.
41. Pinchbeck, 206.

42. Spencely, 'The English Pillow Lace Industry', 70.
43. Pinchbeck, 222.
44. Snell, 'Agricultural Seasonal Unemployment', 434–7.
45. Clapham, *Economic History*, I, 183.
46. Pinchbeck, 225.
47. Pinchbeck, 229, and Clapham, *Economic History*, I, 183.
48. Pollard, *Peaceful Conquest*.
49. Burley; Chapman, 'Industrial Capital'.
50. This paragraph is based on Pollard, *Peaceful Conquest*, 14–16.
51. Honeyman, *Origins of Enterprise*, 28–9.
52. Dickson, 100–16.
53. Beckett, 'Regional Variation', concludes that this depression was
    less intense than previously supposed, and its effects on the rest of
    the economy much weaker. On eighteenth century growth see
    Wrigley, 'The Growth of Population'; Crafts, 'British Economic
    Growth'; and McCloskey, 'The Industrial Revolution'.
54. Cole, 'Factors in Demand 1700–80', 53.
55. This paragraph is based on Little, 65–85.

6. Domestic Manufacture and Women's Work

1. Medick, 'Proto-industrial Family Economy'.
2. Levine, *Family Formation*, 9.
3. ibid, 9, 14.
4. Smith, *Wealth of Nations*, I, 134.
5. Chayanov, *Theory of Peasant Economy*.
6. Harrison, 'Chayanov'.
7. Laslett, 'Family and Household'.
8. MacFarlane, *Origins of English Individualism*, 78.
9. Hufton, 'Women in History', 132.
10. Medick, 'Proto-industrial Family Economy'.
11. Levine, *Family Formation*, 13.
12. Snell, 'Agricultural Seasonal Unemployment'.
13. Hufton, 'Women, Work and Marriage'; Duplessis and Howell,
    'Early Modern Urban Economy'.
14. Houston and Snell, 'Proto-industrialization'.
15. Malcolmson, *Life and Labour*, 42.
16. Neeson, 'Opposition to Enclosure'.
17. Laslett, 'Family and Household', 546.
18. Pinchbeck, *Women Workers*, 129.

19. Eden, *State of the Poor*. Cited in Brown, *Essex at Work*, 14.
20. Spencely, 'English Pillow Lace'.
21. Chambers, 'Rural Domestic Industries', 430.
22. McKendrick, 'Home Demand, Women and Children', 187.
23. B. Collins, 'Proto-industrialization', 132–4.
24. Eden, *State of the Poor*, II, 385, III, 739, 814, 876.
25. Chambers, 'Rural Domestic Industries', 438.
26. Cited in the Hammonds, *The Skilled Laborer*, 145.
27. Eden, *State of the Poor*, III, 821.
28. ibid.
29. Clark, *Working Life of Women*, 111.
30. Cited in Richards, 'Women in the British Economy', 341.
31. Pinchbeck, *Women Workers*, 155.
32. The Hammonds, *The Skilled Laborer*, 152.
33. ibid, 149.
34. Cited in Wadsworth and Mann, *The Cotton Trade*.
35. Baines, *History of the Cotton Manufacture*, 159.
36. Wadsworth and Mann, *The Cotton Trade*, 375.
37. ibid, 301.
38. Richards, 'Women in the British Economy', 343.
39. Clapham, *Economic History*, I, 183.
40. Jones, 'Constraints on Economic Growth'.
41. Snell, 'Agricultural Seasonal Unemployment', 436.
42. Hobsbawm, *Age of Revolution*, 55.
43. Landes, *Unbound Prometheus*, 117.
44. Pollard, 'Labour in the British Economy'.
45. Pinchbeck, *Women Workers*; Alexander, 'Women and the London Trades'.
46. Chapman and Chassagne, *European Textile Printers*, 95, 96, 194.
47. Coleman, 'Growth and Decay: the Case of East Anglia', 120, 123–4.
48. Wadsworth and Mann, *The Cotton Trade*, 285, 325, 332, 336.
49. Boserup, *Women's Role in Economic Development*.
50. Elson and Pearson, 'Nimble Fingers'.
51. McKendrick, 'Home Demand, Women and Children', 186.
52. Prothero, *Artisans and Politics*, 35.
53. Phillips and Taylor, 'Sex and Skill', 82–8.
54. Godelier, 'Work and its Representations'.
55. Medick, 'Proto-industrial Family Economy', 304, 307, 310.
56. Chaytor, 'Household and Kinship: Ryton', 30.
57. Laslett, 'Family and Household', 555.

58. B. Collins, 'Proto-industrialization', 133.
59. Levine, *Family Formation*, 48.
60. J. Anderson, *Observations on National Industry*, I, 39.
61. Cited in Hudson, 'From Manor to Mill', 130.
62. Pinchbeck, *Women Workers*, 126.
63. Chaytor, 'Household and Kinship', 48.
64. O. Harris, 'Households and their Boundaries', 8, 150.
65. Kusamitsu, 'Industrial Revolution and Design', 118.
66. Edwards and Lloyd-Jones, 'Smelser and the Cotton Family', 305.
67. Radcliffe, *New System of Manufacture*.
68. The Hammonds, *The Skilled Laborer*, 162.
69. Reid, 'Decline of Saint Monday', 95.
70. David Sabeen has, however, suggested to me that many of the arts typically associated with housewifery only really came into being in the eighteenth and nineteenth centuries. The finer needlework required in the working of cotton and linen fabrics was not practised by most peasant and working-class households clothed in coarse woollen clothing.

7. Custom and Community in Domestic Manufacture and the Trades

1. Sewell, *Work and Revolution*; Rule, *Experience of Labour*; Prothero, *Artisans and Politics*.
2. Prothero, 35.
3. Malcolmson, 126–33.
4. Prothero, 26.
5. Thomis and Grimmett, *Women in Protest*, 72.
6. ibid, 70.
7. B. Taylor, *Eve and the New Jerusalem*, 90–3.
8. Rule, *Experience of Labour*, 167.
9. Thompson, *Making of English Working Class*, chapter 14, 569–659.
10. ibid, 637.
11. Chambers, 'Rural Domestic Industries'.
12. Jones, 'Constraints on Economic Growth', 427.
13. Rule, 167.
14. Quoted in Malcolmson, 150–1.
15. Hudson, 'Manor to Mill', 143.
16. Gregory, *Transformation of Yorkshire*, 110–11.

17. Charlesworth, *Atlas of Rural Protest*, 50, 54.
18. Neeson, 'Opposition to Enclosure', 62.
19. Charlesworth, 69; Thompson, 'Moral Economy'.
20. Ross, 'Survival Networks'; Whipp, 'Potbank and Unions'.
21. Bertaux-Wiame, 'Life History and Migration', 193.
22. Ross, 'Survival Networks', 10.
23. Chaytor, 'Household and Kinship', 49.
24. O. Harris, 'Households as Natural Units'.
25. Wrightson, 'Household and Kinship', 154; Wrightson, *English Society*, 52.
26. Malcolmson, 118.
27. Bushaway, *By Rite*, 201, 174.
28. See Reddy, 'The Textile Trade'.
29. The most common fraud was the false and short reeling of yarn. The length of a hank in the industry was specified as a fixed number of revolutions or coils on a reel, but the circumference of this reel differed between regions and branches of the industry. False and short reeling was done, then, by using a reel of a shorter circumference than the accepted standard for the area, or by enclosing in each hank a smaller number of threads than the standard. See John Styles, 'Embezzlement'.
30. Heaton, *Yorkshire Woollen and Worsted*, 430.
31. O. Harris, 'Households and their Boundaries', 146–9.
32. Ross, 'Survival Networks', 11, 14.
33. Medick, 'Plebeian Culture', 92.
34. Douglas and Isherwood, *World of Goods*, 11, 202.
35. Ross, 'Survival Networks', 4; Bakker and Talsma, 'Women and Work between the Wars', 80.
36. Eden, *State of the Poor*.
37. Harte, 'Protection and the Linen Manufacture'; Eliot, *Mill on the Floss*, 178.
38. Thirsk, *Economic Policy*, 22–3.
39. McKendrick, 'Home Demand, Women and Children', 197.
40. Smith, *Wealth of Nations*, I, 49.
41. Medick, 'Plebeian Culture', 89.
42. Chapman and Chassagne, *European Textile Printers*, 98.
43. Medick, 'Plebeian Culture', 92, 108.
44. See Berg, Hudson, Sonenscher, 'Introduction', *Manufacture in Town and Country*.
45. Chapman and Chassagne, 98.
46. Reid, 'Decline of Saint Monday', 95.

8. The Economic History of Technological Diffusion

1. Landes, *Unbound Prometheus*, 41, 81, 95, 99, 105, 123.
2. For a lucid, layperson's account of some of the theories of technical change, see Nathan Rosenberg, 'The Direction of Technological Change'. For a good, basic survey of the economic theories and some empirical studies of technical change see Arnold Heertje, *Economics and Technical Change*. N. von Tunzelman, 'Technical Progress', discusses the applications of some of this theory to the British experience before 1860.
3. Habakkuk, *American and British Technology*.
4. Marx, *Capital*, I, 391.
5. David, 'Labour Scarcity'.
6. Habakkuk, 'Labour Scarcity'.
7. Saul, *Technological Change*, 'Introduction'.
8. These arguments are put in Berg, 'Power Loom'.
9. Temin, 'Labour Scarcity in America'.
10. David, 'Labour Scarcity', 33.
11. ibid, 28.
12. McCloskey, *Essays on a Mature Economy*.
13. Von Tunzelman, *Steam Power*; Hyde, *Technological Change*.
14. David, 'The Landscape and the Machine'; Sandberg, *Lancashire in Decline*; Temin, 'Decline of the Steel Industry'.
15. Von Tunzelman, 'Technological Change'.
16. David, 'The Landscape and the Machine'; E. J. T. Collins, 'Harvest Technology'; Roberts, 'Sickles and Scythes'.
17. Sandberg, *Lancashire in Decline*, Gutman, *Work, Culture and Society*, 39. For a critique of this see Lazonick, 'Factor Costs and the Diffusion of Ring Spinning'.
18. Hyde, *Technological Change*.
19. Von Tunzelman, *Steam Power*.
20. Rosenberg, 'Machine Tool Industry'.
21. Rosenberg, 'Economic Development', 'The Direction of Technological Change'.
22. J. R. Harris, 'Industry and Technology', 'Skills, Coal and British Industry'; Mathias, 'Skills and Diffusion'.
23. Marglin, 'What Do Bosses Do?' 'The Power of Knowledge'.
24. Berg, *Technology and Toil*, 'Introduction'.
25. Stone, 'Job Structures in Steel'.
26. Lazonick, 'Class Relations and Capitalist Enterprise'.
27. Samuel, 'Workshop of the World'.

28. Von Tunzelman, *Steam Power*.
29. Bruland, 'Industrial Conflict and Technical Innovation'.
30. Lazonick, 'Self Acting Mule'.
31. Zeitlin, 'Craft Control and Division of Labour'.
32. Price, *Masters, Unions and Men*.
33. These are elaborated in Chapters 9 to 12 below.
34. For example, Gray, *The Labour Aristocracy*.
35. Prothero, *Artisans and Politics*, 15.
36. This perspective has recently been uncovered in disputes over masculinity of the tailoring trades in early nineteenth-century London. See B. Taylor, *Eve and the New Jerusalem*.

9. The Textile Industries: Organizing Work

1. Defoe, *Complete English Tradesman*, 393.
2. Jenkins and Ponting, *British Wool Textiles*, 5.
3. Eden, *State of the Poor*.
4. Jenkins and Ponting, 4, 7.
5. ibid, 58–9.
6. ibid, 75.
7. Heaton, *Yorkshire Woollen and Worsted*, 264–75.
8. R. G. Wilson, 'Supremacy of Yorkshire', 233; Lloyd Prichard, 'Decline of Norwich', 374–6; Wilson, *England's Apprenticeship*, 291.
9. Brown, *Essex*, 2–11, 20.
10. Jenkins and Ponting, 71.
11. Mann, *Cloth Industry in the West of England*, 159–63.
12. Heaton, 78, 285.
13. ibid, 289.
14. Thirsk, 'Fantastical Folly', 62–3.
15. Chambers, 'Rural Domestic Industries', 428–9.
16. Heaton, 235.
17. Rogers, 'Framework Knitting', 8–10; Levine, *Family Formation*, 21.
18. Clapham, *Economic History*, I, 145.
19. Timmins, *Birmingham*, 179–83; Prest, *Coventry*, 53.
20. De Vries, *Economy of Europe*, 100; Harte, 'Rise of Protection', 109.
21. Harte, ibid, 76.
22. ibid, 109.
23. ibid, 103.
24. Bremner, *Industries of Scotland*, 214–30.
25. B. Collins, 'Proto-industrialization', 129–30.

26. Harte, 112.
27. Chapman and Chassagne, 4.
28. Lee, *Cotton Enterprise*, 6, 24.
29. Bremner, 281; Jewkes, 'Localization of Cotton'.
30. Quoted in Bremner, 282.
31. Mann, *Cloth Industry in the West of England*, 97.
32. Heaton, 96, 203.
33. S. C. on Woollen Manufacture, 1806; Heaton, 293–4.
34. Defoe, *Tour*, 491–2.
35. Hudson, 'Proto-industrialization', 51.
36. R. G. Wilson, 238.
37. Jenkins, *West Riding Wool Textiles*.
38. Rogers, 'Framework Knitting', 8–10, 17; Mills, 'Proto-industrialization'.
39. Warner, *Silk Industry*, 499.
40. Aspin and Chapman, *Hargreaves*, 30.
41. ibid, 38.
42. The Hammonds, *Skilled Laborer*, 209.
43. Warner, *Silk Industry*, 513.
44. Prest, 45; Lane, 'Apprenticeship', 316.
45. Prest, 49.
46. Timmins, 179–83.
47. Bremner, 222.
48. Lee, 2.
49. Edwards, *British Cotton Trade*, 9.
50. Lee, 3.
51. Chapman and Chassagne, 37–52.
52. Fitton and Wadsworth, *Strutts and Arkwrights*, 82.
53. Edwards, 131, 145.
54. Friedman, *Industry and Labour*, 151.
55. Jenkins and Ponting.
56. Heaton, 351–2.
57. Hudson, 'Proto-industrialization'.
58. Aspin and Chapman.
59. Mills, 14; Levine, 19.
60. Warner, 58, 128, 139, 154.
61. Prest, 53.
62. Durie, 'Linen Industry', 91.
63. Bremner, 247–8.
64. Edwards and Lloyd-Jones, 'Smelser and the Cotton Factory Family', 306.

65. Wadsworth and Mann, cited in Edwards and Lloyd-Jones, 307.
66. Honeyman, 240.
67. Unwin, *Samuel Oldknow*, 32.
68. Thompson, *The Making of the English Working Class*, 304.
69. ibid, 305.
70. Fitton and Wadsworth, 193. 'The large employer did most of the speaking before the Parliamentary Committees, but he was hardly the characteristic figure of the trade. There were actually very few "cotton lords".'
71. Unwin, *Oldknow*, 15.
72. Lee, 146; Collier, *Family Economy*, 42.
73. Chapman, *Early Factory Masters*, 128; Chapman, 'Fixed Capital in Cotton'. Drinkwater's Manchester mill in 1792 employed 500. See Chaloner, 'Owen, Drinkwater and Factory System', 88. Mills similar in size were the Pleasley Works valued at £4,195, Ribertson's mill at Linby valued at £3,600, Dale's mill at New Lanark valued at £3,500. These were Arkwright licensees and appear to have been restricted to 1,000 spindles. See Chapman, *Early Factory Masters*, 128.
74. Chapman, ibid, 64.
75. Fitton and Wadsworth, 193; Chapman, *Early Factory Masters*, 129.
76. Clapham, 'Factory Statistics', 475–7.
77. Cited in Gatrell, 'Labour, Power and Size', 95.
78. Chapman, 'Fixed Capital in the Industrial Revolution', cited in Gatrell, 109.
79. ibid, 113.
80. Lloyd Jones and Le Roux, 'Size of Firms'.

10. The Textile Industries: Technologies

1. Linebaugh, 'Labour History', 320.
2. Kriedte, Medick, Schlumbohm.
3. Lee, *Cotton Enterprise*, 4.
4. Aspin and Chapman, 57.
5. Jenkins and Ponting, 48; Heaton, 340; and Mann, *Cloth Industry in the West of England*, 126.
6. Lee, *Cotton Enterprise*, 4; Smelser, *Social Change*, 89.
7. Aspin and Chapman, 37–8.
8. Catling, 'Spinning Mule', 39.
9. Edwards, 5 and 8.

10. Von Tunzelman, *Steam Power*, 176–7.
11. *Rees's Manufacturing Industry*, V, 474.
12. Hills, 'Hargreaves, Arkwright and Crompton'.
13. Unwin, *Oldknow*, 32.
14. Von Tunzelman, *Steam Power*, 179–83.
15. Edwards, 204.
16. Lardner, *Silk Manufacture*, 196–200.
17. *Rees's Manufacturing Industry*, IV, 468–9.
18. Ponting, *Woollen Industry in the South West of England*, 61.
19. Heaton, *Yorkshire Woollen and Worsted Industry*, 340.
20. Lardner, 223.
21. Kusamitsu, 53–4.
22. Warner, 117.
23. Timmins, *Birmingham*, 179–89.
24. Baines, *History*, 207; Wood, *History of Wages*, 141–3. In 1815 the few power looms in Stockport were capable only of weaving strong calicoes and other coarse cloth made from low counts of yarn. A greater part of the weaving was done by hand. See Giles, 'Stockport', 37.
25. Wilkinson, 'Power Loom Developments', 129.
26. Baines, *History*, 229–31; McCulloch, 'Rise of Cotton Manufacture'. Also see *P.P.* 1833, Testimony of H. W. Sefton, 622. 'There are now hundreds of dressers, who dress the warp previous to it being woven on the power loom, those are classes who 20 or 30 years ago were scarcely known by name, compared with their present number . . . in one mill, there are about 17 to about 350 looms, that will be one to 20 in Stockport . . . there would be about 500.'
27. The evidence of Babbage quoted by Habakkuk, *American and British Technology*, 148.
28. Contrary to Habakkuk's belief stated ibid, 148.
29. Thompson, *English Working Class*, 315; Wood, *History of Wages*, 141–3.
30. *Rees's Manufacturing Industry*, V, 479.
31. Pinchbeck, *Women Workers*, 149.
32. See Berg, *Machinery Question*, chap. 4.
33. Von Tunzelman, 'Technical Progress', 155, 161.
34. Aspin and Chapman, 48.
35. Von Tunzelman, *Steam Power*, 176.
36. Pinchbeck, 150–1.
37. Scott-Taggart, 'Crompton's Mule', 28.

38. Catling, 43.
39. Baines, *History*, 436.
40. Smout, *History of the Scottish People*, 385–6.
41. Catling, 49.
42. B. Collins, 'Proto-industrialization'.
43. Lee, 24.
44. Unwin, *Oldknow*, 45.
45. *P.P.* 1833, 75.
46. Baines, *History*, 485.
47. Wilkinson, 130.
48. Baines, *History*, 436; Gaskell, *Manufacturing Population*, 241.
49. Baines, 436.
50. *P.P.* 1831, 254.
51. M. Anderson, *Family Structure*, 115; Testimony of G. A. Lee, *P.P.* 1816, 343.
52. Arkwright, *P.P.* 1816, 279; Boyson, 103.
53. Kusamitsu, 118.
54. Chapman and Chassagne, 95, 96, 194.
55. Brown, *Essex*, 20.
56. Mann, *Cloth Industry in the West England*, 126, 149.
57. Randall, 284–5.
58. Heaton, 340; Mann, *Cloth Industry*, 174–92.
59. See above, 121–2. Jones, 'Constraints on Economic Growth'.
60. Aspin and Chapman, 47; Baines, *History*, 159.
61. Aspin and Chapman, 31.
62. Thompson, *English Working Class*, 604–28, 657.

11. The Metal and Hardware Trades

1. Marx, *Capital*, I.
2. Faucher, cited in Clapham, *Economic History of Modern Britain*, I, 175.
3. Cited in Armytage, *History of Engineering*, 93.
4. J. R. Harris, 'Industry and Technology'; Mathias, 'Skills and Innovations'.
5. This paragraph and the passages quoted are based on J. R. Harris, 'Skills, Coal and Industry', 177–9.
6. Cited in Briggs, 'Metals and Imagination', 665.
7. Rolt, *Tools*, 68.
8. Pole, *Fairbairn*, 26, 33.

9. Clapham, *Economic History*, I.
10. Tann, 'Textile Millwright'.
11. Fairbairn, *Mills and Millwork*, vi.
12. Tann, 'Textile Millwright', 82, 87.
13. Jeffreys, 15.
14. More, *Skill and the English Working Class*, ('Early Engineering Workers').
15. Barker and Harris, *St. Helens*; Bailey and Barker, 'Watchmaking in S. W. Lancashire,'; Ashton, *Peter Stubs*, 2–5.
16. Rolt, *Tools*, 68–9.
17. Smiles, *Iron Workers and Tool Makers*, 180.
18. Pole, 39.
19. Musson and Robinson, 'Steam Power'.
20. Armytage, 118, 127.
21. Ashton, *Peter Stubs*, 19.
22. Landes, 'Watchmaking', 11.
23. Roll, *Boulton and Watt*, 18; Clapham, *Economic History*, I, 154.
24. Cited in Briggs, 'Metals', 667–8.
25. Burgess, '1852 Lockout', 218, 221.
26. Allen, *Birmingham and the Black Country*, 17.
27. Hamilton, *Brass and Copper*, 88, 96.
28. Hey, *Rural Metalworkers*, 7, 21.
29. Frost, 'Yeomen and Metalsmiths'.
30. Hay, 'Manufacturers and the Criminal Law', 3.
31. Lloyd, *Cutlery Trades*, 273–7.
32. ibid, 37–50; Hall, 'Trades of Sheffield', 11, 17–18.
33. Lloyd, *Cutlery Trades*, 35, 60.
34. Fairbairn, *Mills and Millwork*, viii.
35. Pole, 92.
36. Roll, *Boulton and Watt*, 225.
37. Pole, 112–17.
38. Armytage, 118–27; Lloyd, 119.
39. Court, *Midland Industries*, 53, 60; Lloyd, 247.
40. Clapham, *Economic History*, I, 173–7.
41. Behagg, 'Custom, Class and Change', 466.
42. Reddy, 'Textile Trade'.
43. Ashton, *Peter Stubs*, 36.
44. Lloyd, 191–7.
45. Donnelly and Baxter, 'Sheffield and the English Revolutionary Tradition', 90–1.
46. Behagg, 464.

47. Lloyd, 203.
48. Ashton, *Peter Stubs*, 28.
49. Hopkins, 'Working Hours', 54–5; Sabel and Zeitlin, 'Historical Alternatives', 44.
50. Court, *Midland Industries*, 218; Allen, *Birmingham and the Black Country*, 152–4.
51. ibid, 152; Reid, 'Decline of Saint Monday', 95–6.
52. Allen, 160, 164.
53. Kelly, 'Brass Trades', 43.
54. Roll, *Boulton and Watt*, 186, 194, 201.
55. Armytage, 126.
56. Pole, 47.
57. Roll, 273.
58. Landes, *Unbound Prometheus*.
59. Burgess, '1852 Lockout', 222; Behagg, 463, 466.
60. Linebaugh, 'The Thanatocracy and Old Mr Gory', outlines a typology of work and modes of production, but fails to analyse any movement or dynamic in these modes.

12. The Birmingham Toy Trades

1. Cited in Eversley, 'Industry and Trade', 87. See *Journals of the House of Commons*, 1759, 496; Robinson, 'Boulton and Fothergill', 61; Eversley, 103; *Victoria History of the Counties of England*, II, *Warwick*, 199, 214.
2. Rowlands, *Masters and Men*, 127.
3. Hamilton, *Brass and Copper*, 131; Lane, 'Apprenticeship', 223.
4. Rowlands, *Masters and Men*, 135. But Taylor and Garbett reported that only 6,000 were thus employed. See *Journals of the House of Commons*, 1759, 496.
5. Lane, 'Apprenticeship', 213.
6. Hamilton, *Brass and Copper*, 267.
7. *House of Commons*, 1759, 497.
8. Eversley, 'Industry and Trade', 94–5; *Victoria History, Warwick*, 214.
9. Hamilton, *Brass and Copper*, 273.
10. Rowlands, *Masters and Men*, 155.
11. ibid, 147, 150; Pelham, 'Immigrant Population'.
12. Court, *Midland Industries*, 218.
13. Hamilton, *Brass and Copper*, 82, 143, 162, 236, 252, 256–8, 264–6; Hutton, *History of Birmingham*, 113.

14. Behagg, 454.
15. *Aris's Gazette*, 26 June 1778, 7 May 1781, 22 April 1782, 26 April 1784, 13 June 1785; *House of Commons, P.P.* 1812.
16. Fitzmaurice, *Life of Shelbourne*, 404.
17. *Victoria History, Warwick*, 238; Eversley, 'Industry and Trade', 96; *Victoria History, Warwick*, 236; Hamilton, *Brass and Copper*, 255. For a more detailed discussion of the manufacturing process in pinmaking see Diderot, *Encyclopédie*.
18. Court, *Midland Industries*, 194; Eversley, 87; Davies, 'Nail Trade', 265.
19. *Victoria History, Warwick*, 238; Eversley, 99.
20. Hamilton, *Brass and Copper*, 266.
21. Sandilands, 'Midlands Glass Industry', 35.
22. Myska, 'Ironmaking in the Czech Lands'.
23. Prosser, *Birmingham Inventors*, for details of the patents, inventions and improvements in the Birmingham trades before 1850. Also see Eversley, 213.
24. See *Aris's Gazette*, 26 Jan., 23 Feb. and 2 March 1761, and 4 July 1763.
25. Hawkes Smith, *Birmingham and Vicinity*, 18.
26. Eversley, 'Industry and Trade', 93.
27. Hawkes Smith *Birmingham and Vicinity*, 11, 13–14.
28. Timmins, *Birmingham and the Midland Hardware District*, 560; Allen, 332.
29. Hawkes Smith, 16.
30. Fitzmaurice, 404.
31. See *Aris's Gazette*. Advertisements of firms selling off shop equipment 1750–1.
32. *Victoria History, Warwick*, 214.
33. Pelham, 'Water Power Crisis', 75–90; Court, 257; Behagg, 'Custom, Class and Change'.
34. Rowlands, *Masters and Men*, 147, 150; Hutton, *Birmingham*, 69.
35. *House of Commons Journals*, 1759, 497; Robinson, 'Boulton and Fothergill'.
36. See the discussion by Hopkins, 'Working Hours', which simply sees this small-scale production as an alternative to industrialization.
37. Disraeli, *Sybil*, 165.
38. Behagg, 'Custom, Class and Change', 458, 464.
39. Hay, 'Manufacturers and the Criminal Law', 47.
40. Church, *Kenricks*, 23.

41. Martin, *Natural History of England* (Warwickshire), 141.
42. Cited in Lane, 'Apprenticeship', 99.
43. *Aris's Gazette*, 15 June 1767.
44. Lane, 'Apprenticeship', 223; Eversley, 110–11.
45. *Aris's Gazette*, 3 Nov. 1777.
46. Lane, 'Apprenticeship', 210.
47. *P.P.* III, 1812, 52. Also see Roche, 'Birmingham Jewellery Trade', 16.
48. Eversley, 110.
49. Rowlands, 82.
50. *Aris's Gazette*, 31 May 1756; 2 April 1764; 28 Dec. 1767; 23 Sept. 1754; 21 March 1791; 28 Dec. 1795.
51. Hay, 'Manufacturers and the Criminal Law', 7–15; *Aris's Gazette*, 6 April 1769 and 13 April 1769.
52. Cited in Porter, *English Society*, 213–14.
53. Eversley, 110–11. Also see Loveridge, 'Wolverhampton Trades', 121–3.
54. *Aris's Gazette*, 1788.
55. Allen, *Industrial History of Birmingham*, 168.
56. Hutton, *Birmingham*.
57. *Aris's Gazette*. Advertisements of trade announcements, 1750–96.
58. Reid, 'Decline of Saint Monday'.

Bibliography

Alexander, S., 'Women and the London Trades', in Mitchell, J. and Oakley, A., *The Rights and Wrongs of Women*. Harmondsworth 1978.

Allen, G. C., *The Industrial History of Birmingham and the Black Country 1860–1927*. London 1929.

Anderson, A., *Historical and Chronological Deduction of the Origin of Commerce from the Earliest Accounts to the Present Times*, 2 vols. London 1764.

Anderson, B. L., 'The Attorney and the Early Capital Market in Lancashire', in Crouzet, F., ed., *Capital Formation in the Industrial Revolution*. London 1972.

Anderson, J., *Observations on the Means of Exciting a Spirit of National Industry, chiefly intended to promote the Agriculture, Commerce, Manufactures and Fisheries of Scotland*, 2 vols. Dublin 1779.

Anderson, M., *Family Structure in 19th Century Lancashire*. Cambridge 1971.

Appleby, J., *Economic Thought and Ideology in Seventeenth Century England*. Princeton 1978.

Aris's Birmingham Gazette (1750–1800).

Armytage, W. H., *A Social History of Engineering*. London 1961.

Ashton, T. S., *An Economic History of England: the 18th Century*. London 1955.

An Eighteenth Century Industrialist, Peter Stubs of Warrington, 1756–1806. Manchester 1939.

Aspin, C. and Chapman, S. C., *James Hargreaves and the Spinning Jenny*. Preston 1964.

Bailey, F. A. and Barker, T. C., 'The Seventeenth Century Origins of Watchmaking in S. W. Lancashire', in Harris, J. R., ed., *Liverpool and Merseyside*. London 1967.

346   *Bibliography*

Baines, E., *A History of the Cotton Manufacture in Great Britain* (1835). London 1966.

Bakker, N. and Talsma, J., 'Women and Work between the Wars: the Amsterdam Seamstresses', in Thompson, P., *Our Common History: the Transformation of Europe*. London 1982.

Barker, T. C. and Harris, J. R., *A Merseyside Town in the Industrial Revolution, St. Helens 1750–1900*. London 1959.

Beckett, J. V., 'Regional Variation and the Agricultural Depression 1730–50', *Economic History Review*, xxxv, Feb. 1982.

Behagg, C., 'Custom, Class and Change in the Trade Societies of Birmingham', *Social History*, iv, Oct. 1979.

Berg, M., 'The Introduction and Diffusion of the Power Loom 1789–1842', M.A. Thesis, University of Sussex 1972.

The Machinery Question and the Making of Political Economy 1815–1848. Cambridge 1980.

'Political Economy and the Principles of Manufacture 1700–1800', in Berg, M., Hudson, P. and Sonenscher, M., *Manufacture in Town and Country before the Factory*. Cambridge 1983.

'Technology and the Age of Manufacture: the Birmingham Toy Trades', in *Workshop on New Perspectives in the History of Technology*. Manchester, March 1981.

Technology and Toil in Nineteenth Century Britain. London 1979.

Berg, M., Hudson, P. and Sonenscher, M., eds., *Manufacture in Town and Country before the Factory*. Cambridge 1983.

Bertaux-Wiame, I., 'The Life History Approach to the Study of Internal Migration: How Women and Men Came to Paris between the Wars', in Thompson, P., *Our Common History: the Transformation of Europe*. London 1982.

Blaug, M., 'The Productivity of Capital in the Lancashire Cotton Industry', *Economic History Review*, xiii, 1961.

Boserup, E., *Women's Role in Economic Development*. London 1970.

Boyson, R., *The Ashworth Cotton Enterprise*. Oxford 1970.

Braverman, H., *Labor and Monopoly Capital*. New York 1974.

Bremner, D., *The Industries of Scotland, their Rise, Progress and Present Condition* (1869). London 1969.

Briggs, A., 'Metals and the Imagination in the Industrial Revolution', *Journal of the Royal Society of Arts*, Sept. 1980.

The Brighton Group, 'The Capitalist Labour Process', *Capital and Class*, I, 1976.

Brown, A. F. J., *Essex at Work 1700–1815*. Chelmsford 1969.

Bruland, T., 'Industrial Conflict as a Source of Technical Innovation: Three Cases', *Economy and Society*, xi (2), 1982.

Burgess, K., 'Technological Change and the 1852 Lockout in the British Engineering Industry', *International Review of Social History*, xiv, 1969.

Burley, K. H., 'An Essex Clothier of the Eighteenth Century', *Economic History Review*, xi, 1958.

Bushaway, B., *By Rite: Custom, Ceremony and Community in England 1700–1880*. London 1982.

Bythell, D., *The Handloom Weavers*. Cambridge 1969.
The Sweated Trades. London 1978.

Catling, H., 'The Development of the Spinning Mule', *Textile History*, ix, 1978.

Cary, J., *An Essay on the State of England* (1695). Bristol 1745.
An Essay Towards Regulating the Trade and Employing the Poor of the Kingdom, 2nd edition. Bristol 1719.

Chaloner, W. H., 'Robert Owen, Peter Drinkwater and the Early Factory System in Manchester, 1788–1800', *Bulletin of the John Rylands Library*, 1954.

Chambers, J. D., 'Enclosure and Labour Supply in the Industrial Revolution', *Economic History Review*, v, 1953.
'The Rural Domestic Industries During the Period of Transition to the Factory System, with Special Reference to the Midland Counties of England', *Proceedings of the Second International Congress of Economic History*, Aix-en-Provence, ii, 1962.
'The Vale of Trent 1660–1800', *Economic History Review Supplement*, no. iii, 1957.

Chapman, S. D., *The Early Factory Masters* Newton Abbot. 1967.
'Fixed Capital in the Cotton Industry', *Economic History Review*, xxiii, 1970.
'Fixed Capital in the Industrial Revolution in Britain', *Journal of Economic History*, xxiv, 3, 1964.

'Industrial Capital Before the Industrial Revolution 1730–1750', in Harte, N. and Ponting, K., *Textile History and Economic History*. Manchester 1973.

Chapman, S. D. and Chassagne, S., *European Textile Printers in the Eighteenth Century: a Study of Peel and Oberkampt*. London 1981.

Charlesworth, A., *An Atlas of Rural Protest in Britain 1548–1900*. London 1983.

Chayanov, A. V., *The Theory of Peasant Economy* (1915), ed. Thorner Kerblay and Smith. Homewood, Ill. 1966.

Chaytor, M., 'Household and Kinship: Ryton in the Late 16th and Early 17th Centuries', *History Workshop*, x, 1980.

Church, R., *Kenricks in Hardware. A Family Business 1791–1966*, Newton Abbot 1969.

Clapham, J. H., *An Economic History of Modern Britain*, 3 vols. Cambridge 1938.

'Some Factory Statistics of 1815–1816', *Economic Journal*, xxv, 1915.

Clark, A., *Working Life of Women in the Seventeenth Century* (1919). London 1982.

Clayre, A., *Nature and Industrialization*. Oxford 1977.

Coates, W. A., 'Changing Attitudes to Labour in the Mid-Eighteenth Century', in Flinn, M. W. and Smout, T. C., eds., *Essays in Social History*. Oxford 1974.

Cole, W. A., 'Factors in Demand', in Floud, R. and McCloskey, D., eds., *The Economic History of Britain since 1700*, vol. 1 1700–1860. Cambridge 1981.

Coleman, D. C., 'Growth and Decay during the Industrial Revolution: the Case of East Anglia', *Scandinavian Economic History Review*, x, no. 2, 1962.

The British Paper Industry, 1495–1860: a Study in Industrial Growth. Oxford 1958.

'Proto-industrialization: a Concept too Many', *Economic History Review*, xxxvi, 1983.

Collier, F., *The Family Economy of the Working Classes in the Cotton Industry*. Manchester 1964.

Collins, B., 'Proto-industrialization and Pre-Famine Emigration', *Social History*, vii, no. 2, 1982.

Collins, E. J. T., 'Harvest Technology and Labour Supply in Britain, 1790–1870', *Economic History Review*, xxii, 1969.

Cooper, J., 'In Search of Agrarian Capitalism', *Past and Present*, lxxx, August 1978.

Court, W. H. B., *The Rise of the Midland Industries 1600–1838*. London 1953.

Crafts, N., 'British Economic Growth, 1700–1831: a Review of the Evidence', *Economic History Review*, xxxvi, 1983.

Cunningham, J., *An Essay on Trade and Commerce*. London 1770.

Cunningham, W., *The Growth of English Industry and Commerce*. Cambridge 1882.

Daniels, G. W., 'Industrial Lancashire Prior to and Subsequent to the Invention of the Mule', *Jrl. of the Textile Institute*, 1927.

Darby, H. C., *A New Historical Geography of England after 1600*. Cambridge 1976.

David, P., 'Labour Scarcity and the Problem of Technological Practice and Progress in 19th Century America', in David, P., *Technical Choice Innovations and Economic Growth: Essays on American and British Experience in the Nineteenth Century*. Cambridge 1975.

'The Landscape and the Machine', in McCloskey, D., ed., *Essays on a Mature Economy: Britain after 1840*. London 1971.

Davies, E. I., 'The Handmade Nail Trade of Birmingham and District', M. Com. Thesis, University of Birmingham, 1933.

Davis, R., 'The Rise of Protection in England 1689–1786', *Economic History Review*, xix, 1966.

Deane, P., *The First Industrial Revolution*. Cambridge 1962.

Deane, P. and Cole, W. A., *British Economic Growth 1688–1959*. Cambridge 1969.

Defoe, D., *A Tour through the Whole Island of Great Britain* (1720). Harmondsworth 1971.

The Complete English Tradesman. London 1726.

Dickson, D., 'Aspects of the Rise and Decline of the Irish Cotton Industry', in Cullen, L. M. and Smout, T. C., *Comparative Aspects of Scottish and Irish Economic and Social History 1600–1900*. Edinburgh 1977.

Diderot, M., *Encyclopédie ou Dictionnaire Raisonné des Sciences, des Arts et des Métiers*. Paris 1755.

'Die Sinking', in Timmins, S., ed., *The Resources, Products and Industrial History of Birmingham and the Midland Hardware District*. London 1866.

Disraeli, B., *Sybil or the Two Nations*. London 1845.
Dobb, M., *Studies in the Development of Capitalism*, 2nd edition. New York 1963.
Donnelly, F. K. and Baxter, J. L., 'Sheffield and the English Revolutionary Tradition 1791–1820', in Pollard, S. and Holmes, C., eds., *Essays in the Economic and Social History of South Yorkshire*. Sheffield 1976.
Douglas, M. and Isherwood, B., *The World of Goods: Towards an Anthropology of Consumption* (1978). Harmondsworth 1980.
DuPlessis, R. and Howell, M.C., 'Reconsidering the Early Modern Urban Economy: the Cases of Leiden and Lille', *Past and Present*, xciv, 1982.
Durie, A. J., 'The Linen Industry in Scotland', in Cullen, D. and Smout, C., *Comparative Aspects of Scottish and Irish Economic and Social History 1600–1900*. Edinburgh 1977.
Eden, F., *The State of the Poor*. 5 vols. (1797). London 1966.
Edwards, M. M., *The Growth of the British Cotton Trade 1780–1815*. Manchester 1967.
Edwards, M. M. and Lloyd-Jones, R., 'N. J. Smelser and the Cotton Factory Family', in Harte, N. B. and Ponting, K. G., eds., *Essays in Textile History*. Manchester 1973.
Eliot, George, *The Mill on the Floss* (1860). London 1902.
Elson, D. and Pearson, R., 'Nimble Fingers Make Cheap Workers', *Feminist Review*, vii, Spring 1981.
Engels, F., *The Condition of the Working Class in England* (1845), ed. E. J. Hobsbawm. London 1969.
Eversley, D. C., 'Industry and Trade 1500–1800', *Victoria County History of Warwickshire*, vii, London 1965.
'The Home Market and Economic Growth in England 1780', in Jones, E. L. and Mingay, G., *Land, Labour and Population in the Industrial Revolution*. London 1967.
Fairbairn, W., *Treatise on Mills and Millwork*. London 1861.
Feinstein, C., 'Capital Formation', in Mathias and Postan, eds., *The Cambridge Economic History of Europe*, vol. vii, part ii. Cambridge 1978.
Fitton, R. S. and Wadsworth, A. P., *The Strutts and the Arkwrights 1758–1830*. Manchester 1958.
Fitzmaurice, Lord Edward, *Life of William, Earl of Shelbourne*, vol. I, 1737–1766. London 1875.

Floud, R. and McCloskey, D., eds., *The Economic History of Britain since 1700*, 2 vols. Cambridge 1981.

Fores, M., 'The Myth of a British Industrial Revolution', *History*, lx, 1981.

'Technical Change and the Technology Myth', *The Scandinavian Economic History Review*, xxx, no. 3, 1982.

Freudenberger, H., 'Proto-industrialization and Protofactories', *Eighth International Congress of Economic History*, Section A–2 Proto-industrialization. Budapest 1982.

Friedman, A., *Industry and Labour*. London 1977.

Froebel, F., Heinrichs, J. and Kreye, O., *The New International Division of Labour*. Cambridge 1980.

Frost, P., 'Yeomen and Metalsmiths: Livestock in the Dual Economy of South Staffordshire, 1560–1720', *Agricultural History*, xxix, 1981.

Gaskell, P., *The Manufacturing Population of England*. London 1833.

Gattrell, V. A. C., 'Labour, Power and the Size of Firms', *Economic History Review*, xxx, 1977.

Gee, J., *The Trade and Navigation of Great Britain Considered*. London 1729.

George, D., *England in Transition*. London 1931.

London Life in the Eighteenth Century. London 1925.

Giles, P. M., 'The Economic and Social Development of Stockport 1815–1836', M.A. Dissertation, Manchester University 1950.

Godelier, M., 'Work and its Representations', *History Workshop*, x, 1980.

Goody, E., *From Craft to Industry. The Ethnography of Proto-industrial Cloth Production*. Cambridge 1982.

Gould, J. D., 'Agricultural Fluctuations and the English Economy in the Eighteenth Century', *Journal of Economic History*, xxii, 1962.

Gray, R. Q., *The Labour Aristocracy in Edinburgh*. London 1980.

Gregory, D., *The Transformation of Yorkshire. A Geography of the Yorkshire Woollen Industry*. London 1982.

Gutman, H. B., *Work, Culture and Society in Industrializing America* (1966). Oxford 1977.

Habakkuk, H. J., *American and British Technology in the Nineteenth Century: the Search for Labour Saving Inventions*. Cambridge 1962.

'Family Structure and Economic Change in Nineteenth Century Europe', *Jrl. of Economic History*, xxv, 1955.

'Labour Scarcity and Technological Change', in Saul, S. B., ed., *Technological Change: the U.S. and Britain in the 19th Century*. London 1970.

'Population Growth in Nineteenth Century Europe', *Jrl. of Economic History*, xvii, 1957.

Hall, J. C., 'The Trades of Sheffield as Influencing Life and Health, More Particularly File Cutters and Grinders', *National Association for the Promotion of Social Sciences*, Oct. 1865.

Hamilton, H., *The English Brass and Copper Industries to 1800*. London 1926.

Hammond, J. L. and Hammond, B., *The Rise of Modern Industry*. London 1925.

The Skilled Laborer 1760–1832 (1919). New York 1970.

Harley, C. K., 'British Industrialization before 1841: Evidence of Slower Growth during the Industrial Revolution', *Jrl. of Economic History*, 1982

Harris, J. R., 'Industry and Technology in the Eighteenth Century: Britain and France', Inaugural Lecture, Birmingham 1971.

'Skills, Coal and British Industry in the 18th Century', *History*, lxi, June 1976.

Harris, O., 'Households and their Boundaries', *History Workshop*, xiii, Spring 1982.

'Households as Natural Units', in Young, K., Wolkowitz, C. and McCullogh, R., *Of Marriage and the Market*. London 1981.

Harrison, M., 'Chayanov and the Economics of the Russian Peasantry', *Jrl. of Peasant Studies*, ii, July 1975.

Harte, N. B., 'The Rise of Protection and the English Linen Trade 1690–1790', in Harte, N. B. and Ponting, K. G., *Textile History and Economic History, Essays in Honour of Julia de Lacy Mann*, Manchester 1973.

Hawkes Smith, W., *Birmingham and its Vicinity as a Manufacturing and Commercial District*. London 1836.

Hay, D., 'Manufacturers and the Criminal Law in the Later Eighteenth Century: Crime and Police in South Staffordshire'. Unpublished Paper presented to the Past and Present Colloquium, Oxford 1983.

Heaton, H., *The Yorkshire Woollen and Worsted Industry*, 2nd edition. Oxford 1965.

Heertje, A., *Economics and Technical Change* (1973). London 1977.

Hey, D., 'A Dual Economy in South Yorkshire', *Agricultural History Review*, 1969.

The Rural Metalworkers of the Sheffield Region, Leicester University Department of English Local History Occasional Papers 2nd Series No. 5, Leicester 1972.

Hills, R. L., 'Hargreaves, Arkwright and Crompton, Why Three Inventors?', *Textile History*, x, 1979.

Hobsbawm, E. J., *The Age of Revolution*, London 1962.

Hohenberg, P., 'Toward a Model of the European Economic System in Proto-industrial Perspective, 1300–1800', *Eighth International Congress of Economic History*, Section A–2 Proto-industrialization. Budapest 1982.

Honeyman, K., *Origins of Enterprise*. Manchester 1981.

Hopkins, E., 'Working Hours and Conditions during the Industrial Revolution: a Reappraisal', *Economic History Review*, xxxv, 1982.

Hoskins, W. G., *The Midland Peasant*. London 1965.

House of Commons, Minutes of Evidence taken before the Committee of the Whole House to Consider Several Petitions Against the Orders in Council, Reports from Committees, *Parliamentary Papers*, iii, 1812.

Houston, R. and Snell, K., 'Proto-industrialization? Cottage Industry, Social Change and the Industrial Revolution', *Historical Journal*, xxvii, 2, 1984.

Howell, C., 'Peasant Inheritance Customs in the Midlands 1280–1800', in Goody, J., Thirsk, J. and Thompson, E. P., *Family and Inheritance, Rural Society in Western Europe 1200–1800*. Cambridge 1976.

Hudson, P., 'From Manor to Mill. The West Riding in Transition', in Berg, M., Hudson, P. and Sonenscher, M.,

Manufacture in Town and Country before the Factory. Cambridge 1983.

'Proto-industrialization: the Case of the West Riding', *History Workshop Journal*, no. 12, Autumn 1981.

Hufton, O., 'Survey Articles. Women in History. I. Early Modern Europe', *Past and Present*, ci, 1983.

'Women, Work and Marriage in Eighteenth Century France', in Outhwaite, R. B., *Marriage and Society*. London 1981.

Hutton, W., *A History of Birmingham to the End of the Year 1780*. Birmingham 1781.

Hyde, C. K., *Technological Change and the British Iron Industry 1700–1870*. Princeton, N.J. 1977.

Jeannin, P., 'La Proto-Industrialisation: Développement ou Impasse?', *Annales Economies, Sociétés, Civilisations*, 35a, 1980.

Jeffreys, J. B., *The Story of the Engineers 1800–1945*. London 1945.

Jenkins, D. T. and Ponting, K. G., *The British Wool Textile Industry 1770–1914*. London 1982.

Jewkes, J., 'The Localization of the Cotton Industry', *Economic History*, ii, no. 5, 1930.

John, A. H., 'Agricultural Productivity and Economic Growth in England 1700–1760', *Journal of Economic History*, xxv, 1965.

Jones, E. L., ed., *Agriculture and the Industrial Revolution*. Oxford 1978.

'Agricultural Origins of Industry', *Past and Present*, 1968.

'Agriculture and Economic Growth in England 1660–1750: Agricultural Change', in Jones, ed., *Agriculture and the Industrial Revolution*. Oxford 1978.

'Agriculture and Economic Growth in England 1650–1815: Economic Change', in Jones, ed., *Agriculture and the Industrial Revolution*. Oxford 1978.

'Environment, Agriculture and Industrialization', *Agricultural History*, li, 1977.

'The Constraints on Economic Growth in Southern England 1650–1850', in *Proceedings of the Third International Congress of Economic History*. Munich 1965.

Journals of the House of Commons, xxxiii, 1759.

Kadish, A., *The Oxford Economists in the Late Nineteenth Century*. Oxford 1982.

Kelley, T., 'Wages and Labour Organization in the Brass Trades of Birmingham and District', PhD. Thesis, Birmingham University 1930.

Kenrick, W., *An Address to the Artists and Manufacturers of Great Britain*. London 1774.

Kriedte P., Medick, H. and Schlumbohm, J., *Industrialization before Industrialisation* (1977), English translation. Cambridge 1981.

Kusamitsu, T., 'The Industrial Revolution and Design', Sheffield University PhD. Thesis, 1982.

Kuznets, S., *Modern Economic Growth*. New Haven and London 1965.

Landes, D., *The Unbound Prometheus. Technological Change and Industrial Development in Western Europe from 1750 to the Present*. Cambridge 1969.

'Watchmaking: a Case Study in Enterprise and Change', *Business History Review*, liii, Spring 1979.

Lane, J., 'Apprenticeship in Warwickshire', vol. ii, University of Birmingham PhD. Thesis, 1977.

Lardner, D., *A Treatise on the Origin, Progressive Improvement and Present State of the Silk Manufacture*. London 1831.

Laslett, P., 'Family and Household as Work Group and Kin Group: Areas of Traditional Europe Compared', in Wall, Robin and Laslett, *Family Forms in Historic Europe*. Cambridge 1983.

Lazonick, W., 'Class Relations and Capitalist Enterprise. A Critical Assessment of the Foundations of Economic Thought.' Howard Economics Research Paper, 1983.

'Factor Costs and the Diffusion of Ring Spinning in Britain Prior to World War 1', *Quarterly Jrl. of Economics*, Feb. 1981.

'Industrial Relations and Technical Change: the Case of the Self Acting Mule', *Cambridge Jrl. of Economics*, iii, 1979.

Lee, C. H., *A Cotton Enterprise 1795–1840. A History of McConnel and Kennedy, Fine Cotton Spinners*. Manchester 1972.

Levine, D., *Family Formation in an Age of Nascent Capitalism*. London 1977.

Lindert, P. H., 'English Occupations, 1670–1811', *Journal of Economic History*, xl (4), December 1980.

'Revising England's Social Tables', *Explorations in Economic History*, 19, 1982.

Lindert, P. H. and Williamson, J. G., 'English Workers' Living Standards during the Industrial Revolution: a New Look', *Economic History Review*, xxxvi, 1983.

Linebaugh, P., 'Labour History Without the Labour Process; a Note on John Gast and his Times', *Social History*, vii, no. 3, 1982.

'The Thanatocracy and Old Mr Gory in the Age of Newton and Locke', mimeo (1983).

Little, A. J., *Deceleration in the 18th Century British Economy*. London 1976.

Lloyd, G. I. H., *The Cutlery Trades*. London 1913.

Lloyd-Jones, R. and Le Roux, A. A., 'The Size of Firms in the Cotton Industry, Manchester, 1815-41', *Economic History Review*. xxxiii, 1980.

Lloyd Prichard, M. F., 'The Decline of Norwich', *Economic istory Review*, 2nd series, ii, 1950.

McCloskey, D. N., ed., *Essays on a Mature Economy. Britain after 1840*. London 1971.

'The Industrial Revolution 1780–1860. A Survey', in Floud, R. and McCloskey, D. N., *The Economic History of Britain since 1700*, vol. 1. Cambridge 1981.

McCulloch, J. R., ed., *Early English Tracts on Commerce* (1856). Cambridge 1952.

'On Cottage and Agrarian Systems', *The Scotsman*, 1 March 1817.

'The Rise, Progress, Present State and Prospects of the British Cotton Manufacture', *Edinburgh Review*, 1827.

MacFarlane, A., *The Origins of English Individualism*. Oxford 1978.

McKendrick, N., 'Home Demand and Economic Growth: a New View of Women and Children in the Industrial Revolution', in McKendrick, ed., *Historical Perspectives, Studies in English Thought and Society*. Cambridge 1974.

Malcolmson, R., *Life and Labour in England 1700–1800*. London 1981.

Maloney, J., 'Marshall, Cunningham and the Emerging Economics Profession', *Economic History Review*, xxiv, 1976.

Mann, J. de Lacy, *The Cloth Industry in the West of England from 1640 to 1880*. Oxford 1971.

Mantoux, P., *The Industrial Revolution in the Eighteenth Century: an Outline of the Beginnings of the Modern Factory System in England*. London 1955.

Marglin, S., 'The Power of Knowledge', in Stephens, F., ed., *Work Organization*. London 1983.

'What Do Bosses Do? The Origins and functions of Hierarchy in Capitalist Production', *Review of Radical Political Economy*, vi, 1974. Version reprinted in Gorz, A., ed., *The Division of Labour*. London 1976.

Marshall, A., *Industry and Trade*, Appendix B. London 1919.

Martin, B., *The Natural History of England or a Description of Each Particular County in Regard to the Curious Productions of Nature and Art*. London 1759.

Martyn, H., 'Considerations on the East India Trade' (1701), in McCulloch, J., ed., *Early English Tracts on Commerce*. Cambridge 1952.

Marx, K., *Capital*, I, Harmondsworth 1976.

The Grundrisse, trans and ed. M. Nicholaus. Harmondsworth 1973.

Mathias, P., *The First Industrial Nation. An Economic History of Britain 1700–1914*. London 1969.

'Skills and the Diffusion of Innovations from Britain in the Eighteenth Century', reprinted in Mathias, P., ed., *The Transformation of England*. London 1979.

The Brewing Industry in England, 1700–1830. Cambridge 1959.

Medick, H., 'Plebeian Culture in the Transition to Capitalism', in Samuel, R. and Stedman Jones, G., *Culture, Ideology and Politics*. London 1982.

'The Proto-Industrial Family Economy', *Social History*, Oct. 1976.

Mendels, F., 'Proto-industrialization: the First Phase of the Process of Industrialization', *Journal of Economic History*, xxxii, 1972.

Mendels, F. and Deyon, P., 'Proto-industrialization: Theory and Reality', *Eighth International Congress of Economic History*, Section A–2 Proto-industrialization. Budapest 1982.

Mill, J. S., *Principles of Political Economy* (1848), in Mill, J. S., *Collected Works*. Toronto 1965.

'The Nature, Origins, and Progress of Rent' (1828), in *Essays on Economics and Society*, Mill, *Collected Works*, iv. Toronto 1967.

Mills, D. R., 'Proto-industrialization and Social Structure: the Case of the Hosiery Industry in Leicestershire, England', *Eighth International Congress of Economic History*, Section A–2. Budapest, 1982.

Millward, R., 'The Emergence of Wage Labour in Early Modern England', *Explorations in Economic History*, 18, 1981.

Mitchell, B. R. and Deane, P., *Abstract of British Historical Statistics*. Cambridge 1962.

More, C., *Skill and the English Working Class*. London 1982.

Musson, A. E. and Robinson, 'The Early Growth of Steam Power', *Economic History Review*, xi, 1959.

Myska, M., 'Pre-industrial Ironmaking in the Czech Lands: the Labour Force and Production Relations c. 1350–1840', *Past and Present*, lxxxii, Feb. 1979.

Neeson, J., 'Opposition to Enclosure in Northamptonshire c. 1760–1800', in Charlesworth, A., *An Atlas of Rural Protest in Britain 1548–1900*. London 1983.

O'Brien, P. K. and Keyder, C., *Economic Growth in Britain and France 1780–1914*. London 1976.

Pearson, R., 'Reflections on Proto-industrialization from a Contemporary Perspective'. Unpublished Paper, 1980.

Pelham, R. A., 'The Immigrant Population of Birmingham 1686–1726', *Birmingham Archaeological Transactions*, lxi, 1937.

'The Water Power Crisis in Birmingham in the Eighteenth Century', *University of Birmingham Historical Journal*, ix, 1963–4.

Phillips, A. and Taylor, B., 'Sex and Skill: Notes Towards Feminist Economics', *Feminist Review*, vi, 1980.

Pinchbeck, I., *Women Workers and the Industrial Revolution* (1930). London 1981.

Polanyi, K., *The Great Transformation*. London 1957.

Pole, W. A., ed., *The Life of William Fairbairn*. London 1877.

Pollard, S., 'The Factory Village in the Industrial Revolution', *English Historical Review*, lxxix, July 1964.

'Industrialization and the European Economy', *Economic History Review*, xxvi, Nov. 1973.

'Labour in the British Economy', *Cambridge Economic History of Europe*, vii, part ii, vol. 1. Cambridge 1978.

Peaceful Conquest. Oxford 1981.

The Genesis of Modern Management. Harmondsworth 1965.

Ponting, K., *The Woollen Industry in the South West of England*. Wiltshire 1971.

Porter, Roy, *English Society in the Eighteenth Century*. Harmondsworth 1982.

Postlethwayt, M., *Britain's Commercial Interests*, vol. 2. London 1757.

Prest, J., *The Industrial Revolution in Coventry*. London 1960.

Price, R., *Masters, Unions and Men. Work Control in Building and the Rise of Labour 1830–1914*. Cambridge 1980.

Prosser, R. B., *Birmingham Inventors and Inventions*. Birmingham 1881.

Prothero, I., *Artisans and Politics in Early Nineteenth Century London. John Gast and his Times*. London 1979.

Radcliffe, W., *Origin of the New System of Manufacture*. Stockport 1828.

Randall, A. J., 'The Shearmen and the Wiltshire Outrages of 1802: Trade Unionism and Industrial Violence', *Social History*, vol. vii, no. 3, 1982.

Reddy, W., 'Skeins, Scales, Discounts, Steam and Other Objects of Crowd Justice in Early French Textile Mills', *Comparative Studies in Society and History*, xxi, 1979.

'The Textile Trade and the Language of the Crowd at Rouen 1752–1791', *Past and Present*, lxxiv, Feb. 1977.

Rees's Manufacturing Industry 1819–20. A Selection from the Cyclopedia; or Universal Dictionary of Arts, Sciences and Literature, by Abraham Rees, edited by Neil Cossons, 5 vols. Newton Abbot 1975.

Reflections on Various Subjects Relating to Arts and Commerce, Particularly the Consequences of Admitting Foreign Artists on Easier Terms. London 1752.

Reid, D., 'The Decline of Saint Monday', *Past and Present*, lxxi, 1976.

'Report and Minutes of Evidence on the State of the Woollen Manufacture of England', *Parliamentry Papers*, 1806.

Richards, E., 'Women in the British Economy', *History*, liii, 1974.

Roberts, M., 'Sickles and Scythes: Women's Work and Men's Work at Harvest Time', *History Workshop Journal*, vii, 1979.

Robinson, E., 'Boulton and Fothergill 1762–1782 and the Birmingham Export of Hardware', *University of Birmingham Historical Journal*, vii, 1959–60.

Rogers, A., 'Rural Industrial and Social Structure: the Framework Knitting Industry of South Nottinghamshire, 1670–1840', *Textile History*, xii, 1981.

Roll, E., *An Early Experiment in Industrial Organization, Being a History of the Firm of Boulton and Watt 1775–1805*. London 1930.

Rolt, L. T. C., *Tools for the Job. A Short History of Machine Tools*. London 1965.

Rosenberg, N., 'Economic Development and the Transfer of Technology: Some Historical Perspectives', *Technology and Culture*, xi, 1970.

'Technological Change in the Machine Tool Industry, 1840–1910', *Journal of Economic History*, xxiii, Dec. 1963.

'The Direction of Technological Change: Inducement Mechanisms and Focusing Devices', *Economic Development and Cultural Change*, xx, 1969.

Ross, E., 'Survival Networks: Women's Neighbourhood Sharing in London before World War One', *History Workshop Journal*, xv, Spring 1983.

Rowlands, M., *Masters and Men in the West Midlands Metalware Trades Before the Industrial Revolution*. Manchester 1975.

'Society and Industry in the West Midlands at the End of the 17th Century', *Midland History*, iv, Spring 1977.

Rule, J., *The Experience of Labour in 18th Century Industry*. London 1981.

Sabel, C. and Zeitlin, J., 'Historical Alternatives to Mass Production', mimeo (1982).

Samuel, R., 'The Workshop of the World: Steam Power and Hand Technology in mid-Victorian Britain', *History Workshop Journal*, iii, Spring 1977.

Sandberg, L., *Lancashire in Decline*. Columbus, Ohio 1974.

Sandilands, D. N., 'The History of the Midlands Glass Industry', M. Com. Thesis, University of Birmingham 1927.

Saul, S. B., 'Editor's Introduction', in Saul, ed., *Technological Change: the U.S. and Britain in the 19th Century*. London 1970.

Schmiechen, J. A., *Sweated Industries and Sweated Labour*. London 1984.

Schmitz, H., *Manufacturing in the Backyard, Case Studies of Accumulation and Employment in Small Scale Brazilian Industry*. London 1982.

Schremmer, E., 'Proto-industrialization', *Journal of European Economic History*, 1982.

Scott-Taggart, W., 'Crompton's Invention and Subsequent Development of the Mule', *Journal of the Textile Institute*, xviii, 1927.

Select Committee on Children in Factories, *Parliamentary Papers*, 1831.

Select Committee on Manufactures, Commerce and Shipping, *Parliamentary Papers*, 1833.

Select Committee on the Employment of Children, *Parliamentary Papers*, 1816.

Sewell, W., *Work and Revolution in Nineteenth Century France*. Cambridge 1982.

Sider, G. M., 'Christmas Mumming and the New Year in Outport Newfoundland', *Past and Present*, lxxi, 1976.

Smelser, N. J., *Social Change in the Industrial Revolution: an Application of Theory to the Lancashire Cotton Industry 1770–1840*. London 1959.

Smiles, S., *Industrial Biography: Iron Makers and Tool Makers*. London 1863.

Smith, A., *An Inquiry into the Nature and Causes of the Wealth of Nations* (1776), 2 vols. Oxford 1976.

Smout, T. C., *A History of the Scottish People 1560–1830*. London 1969.

Snell, K., 'Agricultural Seasonal Unemployment, the Standard of Living and Women's Work in the South and the East 1690–1860', *Economic History Review*, xxxiv, 1983.

Spencely, G. F. R., 'The English Pillow Lace Industry

1845–80: a Rural Industry in Competition with Machinery', *Business History*, lxx, 1970.

Spufford, M., *Contrasting Communities: English Villages in the 16th and 17th Centuries*. Cambridge 1974.

'Peasant Inheritance Customs and Land Distribution', in Goody, J., Thirsk, J. and Thompson, E. P., *Family and Inheritance, Rural Society in Western Europe 1200–1800*. Cambridge 1976.

Stewart, D., *Lectures on Political Economy*, vol. 1 in Hamilton, W., *Collected Works of Dugald Stewart*, viii. Edinburgh 1855.

Stone, K., 'The Origins of Job Structures in the Steel Industry', *Review of Radical Political Economy*, vi, 1976.

Styles, John, 'Embezzlement, Industry and the Law in England, 1500–1800' in Berg *et al.*, eds., *op. cit.*

Tann, J., 'The Textile Millwright in the Early Industrial Revolution', *Textile History*, v, Oct. 1974.

Taylor, A. J., 'Concentration and Specialization in the Lancashire Cotton Industry, 1825–50', *Economic History Review*, i, 1949.

Taylor, B., 'The Men Are as Bad as Their Masters', in Taylor, B., *Eve and the New Jerusalem*. London 1983.

Temin, P., 'The Relative Decline of the British Steel Industry, 1880–1914', in Rosovsky, H., ed., *Industrialization in Two Systems. Essays in Honour of Alexander Gershenkron*. New York 1966.

'Labour Scarcity in America', *Journal of Interdisciplinary History*, I (2), Winter 1971.

Thirsk, J., *Economic Policy and Projects: the Development of Consumer Society in Early Modern England*. Oxford 1978.

'Industries in the Countryside', in Fisher, F. J., ed., *Essays in the Economic and Social History of Tudor and Stuart England*. Cambridge 1961.

The Agrarian History of England and Wales, iv 1500–1640. Cambridge 1967.

'The Fantastical Folly of Fashion: the English Stocking Knitting Industry 1500–1700', in Harte, N. B. and Ponting, K. G., eds., *Textile History and Economic History*. Manchester 1973.

Thomis, M. I. and Grimmett, J., *Women in Protest 1800–1850*. London 1982.

Thompson, E. P., 'The Grid of Inheritance', in Goody, J., Thirsk, J. and Thompson, E. P., *Family and Inheritance, Rural Society in Western Europe 1200–1800*. Cambridge 1976.

The Making of the English Working Class. Harmondsworth 1968.

'The Moral Economy of the Crowd', *Past and Present*, i, 1972.

'Time, Work Discipline and Industrial Capitalism', *Past and Present*, 1967.

Thomson, J., 'Variations in Industrial Structure in Pre-Industrial Languedoc', in Berg, M., Hudson, P. and Sonenscher, M., *Manufacture in Town and Country before the Factory*. Cambridge 1983.

Thorold Rogers, James, E., *Six Centuries of Work and Wages* (1884). London 1917.

Timmins, S., ed., *The Resources, Products and Industrial History of Birmingham and the Midland Hardware District*. London 1866.

Toynbee, A., *Lectures on the Industrial Revolution in England* (1884). London 1969.

Tucker, J., *A Brief Essay on the Advantages and Disadvantages which Respectively Attend France and Great Britain with Regard to Trade*. London 1749.

Instructions for Travellers. London 1757.

Unwin, G., *Samuel Oldknow and the Arkwrights*. Manchester 1924.

The Guilds and Companies of London. London 1908.

Usher, A. P., *An Introduction to the Industrial History of England*. London 1921.

The Victoria History of the Counties of England, ii, *The History of the County of Warwick*. London 1965.

Von Tunzelman, G. N., *Steam Power and British Industrialization to 1860*. Oxford 1978.

'Technical Progress', in Floud and McCloskey, *An Economic History of Britain since 1700*, vol. 1. Cambridge 1981.

Vries, J. de., *The Economy of Europe in the Age of Crisis*. Cambridge 1976.

Wadsworth, A. P. and Mann, J. de L., *The Cotton Trade and Industrial Lancashire 1600–1780*. Manchester 1931.

Warner, F., *The Silk Industry of the U.K.* London 1921.

Westerfield, R. B., *Middlemen in English Business* (1915). London 1968.

Whipp, R., 'Potbank and Unions: a Study of Work and Trade Unionism in the British Pottery Industry', Unpublished PhD. Thesis, University of Warwick 1983.

Wilkinson, W., 'Power Loom Developments', *Official Record of the Annual Conference of the Textile Institute*. Bolton 1927.

Williamson, J. G., 'Why Was British Growth So Slow during the Industrial Revolution?', *Journal of Economic History*, 44 (3), 1984.

Williamson, O. E., 'The Evolution of Hierarchy: an Essay on the Organization of Work', University of Pennsylvania Discussion Paper, Philadelphia 1976.

Wilson, C., *England's Apprenticeship 1603–1763*. London 1965.
'The Other Face of Mercantilism', in Coleman, D. C., ed., *Revisions in Mercantilism*. London 1969.

Wilson, C. and Parker, G., *An Introduction to the Sources of European Economic History 1500–1800*. London 1977.

Wilson, R. G., 'The Supremacy of the Yorkshire Cloth Industry in the Eighteenth Century', in Harte, N. B. and Ponting, K. G., *Textile History and Economic History*. Manchester 1973.

Wood, G. H., *The History of Wages in the Cotton Trade*. London 1910.

Wrightson, K., *English Society 1500–1700*. London 1982.
'Household and Kinship in Sixteenth Century England', *History Workshop Journal*, xii, Autumn 1981.

Wrigley, E. A., 'A Simple Model of London's Importance in Changing English Society and Economy 1650–1750', in Patten, J., ed., *Pre-Industrial England*. Folkestone, Kent 1979.
'The Growth of Population in Eighteenth Century England: a Conundrum Resolved', *Past and Present*, xcviii, 1983.

Zeitlin, J., 'Craft Control and the Division of Labour: Engineers and Compositors in Britain 1890–1930', *Cambridge Jrl. of Economics*, 3, 1979.

Index

Bramah, machine shop of, 41, 279
Brass trades, 27, 52, 84, 265, 275, 283, 291–3
Braverman, Harry, 190–1, 192
British industry, labour intensity of, 182, 184, 254
Building industry, 26–30, 38–9, 195, 315, 318
Buttonmaking, 118, 124, 137, 145–6, 157, 214, 283, 289, 301, 305, 307, 309
By-employment, 54, 102–3
 in rural industry, 104–5, 111, 132
 geographical dispersal, 108
 in textile industry, 211

Calico printing, 76, 84, 172, 279
 rural industry, 105, 147–9, 152
 riots against, 115
 description of, 205, 216–17
 mechanization of, 251–2
 women employed in, 260
Cambridge population data, 100
Canal networks, 126
Capital (Marx), 73, 293
Capital
 accumulation/formation, 15, 50, 78, 86–7, 114–15, 117
 investment, 16
 agriculture a source of, 94, 100
 transfer into land from manufacturing, 117
 restrictions on, 118
 supplies of, 180, 281
 monopolies, 190
 fixed costs, 245
 of Birmingham manufacturers, 290
Capital-intensive production, 182–3, 188
Capitalism
 origins of, 18
 transition to industrial, 70, 81, 131, 304
 private property, 72–3
 and rural production, 74–5, 133
 expansion of, 146
 and luxury expenditure, 171
 linked to community, 173–4

Marxist theory of, 190–3
 range of development, 195–7
 structures of work organization, 219
Carding engines, 31, 52, 121, 145, 221–2, 237, 240, 243–4, 261
 by hand, 255
Carpet weaving, decline in, 112
Cartwright, Edmund, 126
Cary, John, 51
Chapman, S.D., 100, 217
Chapman and Chassagne, 205
Charcoal smelting, 36
Chayanov, A.V., 133
Cheadle Hulme cotton factory, 226
Child labour, 19, 52, 75
 contributions to family income, 100, 122, 132, 135, 137
 in textile industry, 149, 158, 212–13, 224, 244, 259–60
 productivity of, 248, 254–5
 in toy trades, 294, 304, 307, 310–11
Children, training of, 156
Civil War, 109
Clapham, J.H., 19, 23–4, 85–6, 145, 280
Clark, Alice, 19, 136, 142
Clarke, Thomas, 293
Class conflict, 120
Cloth industries, *see* textile industries
Cloth Workers' Company, dissolution of, 114
Clothiers, 52, 53, 54
 replaced by factories, 116, 221–2
 credit for, 118–19
 status of, 162–3, 208
 cooperatives of, 229
Coal, 29, 35, 39, 52, 186
 mining technologies, 107
 in metal manufacture, 266–7
 see also power, sources of
Coalbrookdale ironworks, 36, 269
Cobden, Richard, 268
Coke of Norfolk, 92
Coke smelting, 36, 37, 186
Collins, Brenda, 138
Colquhoun, Patrick, 29, 236
Combing process, 242–3